Chronology of KSC and KSC Related Events for 1999

Elaine E. Liston, InDyne, Inc., Kennedy Space Center, Florida

NASA/TM-2000-208588

February 2000

FOREWORD

This 1999 Chronology is published to describe and document KSC's role in NASA's progress.

Materials for this Chronology were selected from a number of published sources. The document records KSC events of interest to historians and other researchers. Arrangement is by date of occurrence, though the source cited may be dated one or more days after the event.

Materials were researched and prepared for publication by Archivist Elaine E. Liston.

Comment on the Chronology should be directed to the John F. Kennedy Space Center, Archives, LIBRARY-E, Kennedy Space Center, Florida, 32899. The Archivist may also be reached by e-mail at Elaine.Liston-1@ksc.nasa.gov, or (321) 867-2407.

Table of Contents

JANUARY

JANUARY 1: Florida's commercial spaceport here is booked to near capacity next year, solidifying its position in a lucrative field. The reasons for the brisk business include a new rocket, the opening of another launch complex and the first test program of a new space vehicle from Kennedy Space Center in almost 20 years. "The year 1999 is going to be a great year. And the year 2000 is going to be spectacular," said Ed O'Connor, executive director of Spaceport Florida Authority, a state agency working to attract aerospace business. Twenty-six of the missions will be aboard unmanned rockets flown from Cape Canaveral Air Station. The vehicles will carry everything from communications satellites to NASA deep-space probes. An experimental NASA space plane called the X-34 also is to begin flying from KSC in December – the first time a totally new rocket has taken off from there since the shuttle program began in 1981. The small craft will be launched from under the wing of an airplane. It is designed to ferry small payloads into space. Beyond that, the state is continuing its push to lure a next-generation spaceship called VentureStar that may hold the key to 21st century dominance in the world's commercial launch industry. ["Commercial space-launch industry takes off in Florida," <u>Florida Today</u>, January 1, 1999, p 1A & 7A.]

JANUARY 3: Mars Polar Lander Mission Status Report, Sunday, January 3, 1999. Mars Polar Lander - due to become the first spacecraft to set down near the edge of Mars' southern polar cap -- pierced through a blustery, cloud-covered Florida sky at 3:21 p.m. Eastern Standard Time today atop a Delta II launch vehicle from Cape Canaveral Air Station's Launch Complex 17B. The spacecraft, launched successfully on the first day of the launch period, is equipped with a robotic arm to dig beneath the layered terrain of the Martian polar region and two microprobes to crash into the planet's surface and conduct two days of soil and water experiments up to 1 meter (3 feet) below the Martian surface. Sixty-six seconds after liftoff, the Delta's four solid-rocket strap-on boosters were jettisoned. Firing of the main first-stage engine lasted approximately 4 minutes, 24 seconds. Eight seconds later, the first stage was discarded, and 5.5 seconds later the second stage ignited. Four and a half seconds after that, the nose cone surrounding the lander was jettisoned. The second-stage burn lasted 6 minutes, 44 seconds and placed the spacecraft into a low-Earth orbit at an altitude of 191 kilometers (119 miles). The spacecraft coasted over the Indian Ocean for approximately 23 minutes before the second stage engine fired briefly a second time. The third stage fired for 88 seconds at 3:57 p.m. EST to propel the spacecraft out of Earth's gravity and on its way to Mars. At 4:03 p.m. EST, Mars Polar Lander separated from the third stage. A set of solar panels located on the spacecraft's outer cruise stage were deployed shortly thereafter and pointed at the Sun. At 4:19 p.m. EST, the lander's signal was acquired by a 34-meter-diameter (112-foot) antenna of NASA's Deep Space Network in Canberra, Australia. Mars Polar Lander's interplanetary cruise to Mars will take it more than 180 degrees around the Sun in what is called a Type 2 trajectory, allowing the spacecraft to target a landing zone close to Mars' south pole at 73 to 76 degrees south latitude. Throughout the cruise, the

1

spacecraft will communicate with Earth using its X-band transmitter and medium-gain horn antenna mounted on the cruise stage. During the first 30 days of flight, the spacecraft will be tracked 10 to 12 hours per day. Quiet phases of the trip will require only four hours of tracking time each day. The spacecraft is scheduled to fire its thrusters in a trajectory correction maneuver January 18. That maneuver is designed to remove a targeting bias intended to prevent the third stage of the Delta II rocket from following in the lander's flight path and colliding with Mars, as well as any small launch injection errors. That maneuver is expected to take approximately 5 minutes to execute. Mars Polar Lander is the second of two spacecraft launched to the red planet during the December 1998-January 1999 Mars launch opportunity. Mars Climate Orbiter was launched December 11, and is scheduled to reach Mars next September 23. Onboard Mars Polar Lander are two microprobes developed as the Deep Space 2 project under NASA's New Millennium Program. The Deep Space 2 probes will smash into the Martian surface as a test of new technologies for future planetary descent probes. Bruce Buckingham. (1999). **Mars Polar Lander Mission Status Report** [Online]. Available E-mail: domo@news.ksc.nasa.gov/subscribe shuttle-status [1999, January 3].]

JANUARY 4: Two water-hunting spacecraft are sailing smoothly toward Mars today, where they are to arrive later this year to continue NASA's latest exploration of the alluring planet. Launched three weeks apart, the Mars Polar Lander and Mars Climate Orbiter spacecraft are in excellent health as they head into deep space, NASA officials said Monday. ["Spacecraft 'well-behaved' on way to Mars," **Florida Today**, January 5, 1999, p 1B.]

JANUARY 5: On Feb. 6, a Boeing Delta II rocket carrying the Stardust spacecraft is scheduled to launch. Stacking of the rocket at Launch Complex 17A, Cape Canaveral Air Station, begins Tuesday, Jan. 5. Stardust will capture comet particles flying off the nucleus of comet Wild 2 in January 2004, plus collect interstellar dust for later analysis. The collected samples will return to Earth in a re-entry capsule to be jettisoned from Stardust as it swings by in January 2006. ["The Next NASA Payload Launch," **KSC Countdown**, January 5, 1999.]

JANUARY 7: Space Shuttle Status Report, Thursday, January 7, 1999. STS-93: Chandra X-ray Observatory (formerly AXAF). Columbia's payload bay doors were opened earlier this week following the holiday down period. Also this week, workers replaced a transducer for orbiter freon coolant loop No. 2 and testing of the new component is complete. Columbia's forward and midbody compartment closeouts are in work. Potable water sampling is in progress. In the Vehicle Assembly Building, external tank and solid rocket booster closeouts continue. STS-96: 2nd U.S. International Space Station Flight Spacehab Double Module. Processing of the orbiter Discovery has resumed following the holiday down period. Functional testing of the orbiter's purge vent drain system is complete. A corrosion control modification on the base heat shield is in progress and main propulsion system leak and functional testing is under way. Technicians are preparing for upcoming radiator functional tests.

Replacement of Discovery's window No. 1 continues and window No. 7 replacement begins later this week. STS-101: 3rd International Space Station Flight Spacehab Double Module / ICC. Atlantis remains in VAB high bay 2 awaiting Columbia's move to the VAB in early February. Workers are preparing for the installation of Atlantis' onboard tactical air command and navigation system (TACAN). OPF bay 3 will undergo planned open bay work for one week following Columbia's departure and will then house Atlantis throughout the STS-101 processing flow. STS-99: Shuttle Radar Topography Mission (SRTM). Endeavour's orbiter maneuvering system functional tests are in work. Shuttle main engine inspections continue and post flight servicing of the onboard waste management system is in progress. Bruce Buckingham. (1999). **Kennedy Space Center Space Shuttle Status Report** [Online]. Available E-mail: domo@news.ksc.nasa.gov/subscribe shuttle-status [1999, January 7].]

◆ A Japanese communications satellite won't be ready for its scheduled launch next week aboard a Lockheed Martin Atlas rocket, officials said Thursday. The JCSAT-6 spacecraft was slated for a ride into Earth orbit Jan. 13 from Cape Canaveral Air Station. Instead, a problem will force workers to remove the satellite from atop the rocket and return it to a processing facility in Titusville. A spokeswoman for Hughes Space and Communications, which built the spacecraft, could not give further details on the problem. The delay is expected to last a least several weeks. The spacecraft had been scheduled to launch last summer but was grounded after being damaged in an apparent power surge caused by a lightning strike near its processing building. ["Atlas launch of satellite postponed," <u>Florida Today</u>, January 8, 1999, p 7A.]

JANUARY 9: With two pieces of NASA's International Space Station already assembled in orbit, preparations are underway on the next station segments for launch this year from Kennedy Space Center. The parts are to be ferried into orbit on four shuttle flights, beginning with a mission on Discovery in May. Though months away from flight, a large-scale effort to get the pieces ready is unfolding at KSC. Inside a huge KSC building, technicians are working on everything from massive solar arrays to tiny electronics boxes. Altogether, the station will need 33 more shuttle flights and nine Russian rocket launches to carry about 100 major pieces into orbit. ["NASA pieces together station parts," <u>Florida Today</u>, January 10, 1999, p 1A & 2A.]

JANUARY 11: Space Shuttle Status Report, Monday, January 11, 1999. STS-93: Chandra X-ray Observatory (formerly AXAF). Columbia's forward and midbody compartment closeouts continue. Technicians will cycle the left-hand payload bay door today to check recently replaced seals. Managers plan to close the orbiter's payload bay doors Friday in preparation for an early February transfer to the Vehicle Assembly Building. Leak checks on two of the orbiter's hypergolic line disconnects are in work today and this week Ku-band system stowage is scheduled. STS-96: 2nd U.S. International Space Station Flight Spacehab Double Module. Removal of Discovery's caution and warning box is complete. The orbiter's ammonia controller has been replaced and the forward reaction control system (FRCS) has arrived in the OPF for installation tomorrow. Last week, radiator functional tests and inspections revealed a probable micrometeoroid ding on a portion of the radiator and engineering evaluation is now under way. Main propulsion system functional tests are ongoing and corrosion control modifications on the engine heat shields continue. Replacement of Discovery's window No. 1 is complete and window No. 7 replacement is in work today. Workers will evaluate window No. 5 this week. STS-101: 3rd International Space Station Flight Spacehab Double Module / ICC. Atlantis remains in VAB high bay 2 awaiting Columbia's move out of OPF bay 3 in early February. Workers are preparing for the installation of Atlantis' onboard tactical air command and navigation system (TACAN). OPF bay 3 will undergo planned open bay work for one week following Columbia's departure and will then house Atlantis throughout the STS-101 processing flow. STS-99: Shuttle Radar Topography Mission (SRTM). Endeavour's left orbiter maneuvering system (OMS) pod functional tests are complete and good. Functional tests on the right OMS pod are in work. Technicians completed post flight deservicing of the orbiter's waste management system last week. Shuttle main engine inspections continue and preparations for FRCS testing are under way. Bruce Buckingham. (1999). **Kennedy Space Center Space Shuttle Status Report** [Online]. Available E-mail: domo@news.ksc.nasa.gov/subscribe shuttle-status [1999, January 11].]

JANUARY 12: Attendance at Kennedy Space Center Visitor Complex during 1998 increased more than 50,000, from 2.7 million to 2.75 million. Factors contributing to the increased attendance include major additions such as the International Space Station Center and LC 39 Observation Gantry; hosting of space-related major motion picture premiers held at the Visitor Complex; the highly publicized STS-95/John Glenn launch and the launch of the first component of the International Space Station, officials said. Overall Christmas week attendance was up 2 percent over 1997, with 1717, 505 guests visiting the complex in 1998, said Rick Abramson, president of Delaware North Parks Services of Spaceport Inc., which manages the attraction. ["KSC Visitor Complex traffic up," <u>Florida Today</u>, January 13, 1999, p 1C.]

◆ An Internet site for selling and trading items has pulled a listing for a piece of wreckage from NASA's space shuttle Challenger disaster. The listing for an "authentic Challenger O-ring" was removed Tuesday, after officials at eBay Inc. of San Jose, Calif., learned it is illegal to possess debris from the shuttle disaster that killed seven astronauts in 1986. "Any piece or part from Challenger is considered government property," said Jennifer McCarter, spokeswoman at NASA headquarters in Washington, D. C. "If it is indeed authentic, it would be subject to seizure." Federal law allows up to a $10,000 fine, 10 years in prison or both for possession of Challenger wreckage. People who falsely try to sell Challenger debris could be subject to fraud charges. The eBay web site allows people to post listings for items they wish to sell or trade, but the site's policy prohibits the sale of illegal items. Spokeswoman Kristin Seuell said the company would cooperate with law enforcement or NASA if asked to provide information on the seller. ["Web site yanks ad for shuttle wreckage," <u>Florida Today</u>, January 13, 1999, p 1B.]

◆ The Air Force wants to radically change the safety system that has protected Florida from exploding rockets for nearly 50 years, and the proposal is triggering alarm among federal officials. Aerospace experts say the move could endanger people and property along Florida's East Coast as several unproven new rockets start making inaugural flights from cape Canaveral Air Station during the next few years. At issue is who will certify the design of rockets lifting off from the Air Station and NASA's nearby Kennedy Space Center. The safety responsibility currently is shouldered by civilian and military engineers who work for the Air Force as an independent unit at the air base and live in surrounding towns most exposed to potential danger during launch accidents. To save money, the Air Force wants to take control of the work and transfer it to Air Force Materiel Command at Wright-Patterson Air Force Base in Dayton, Ohio. The state of Florida wants to increase the number of launches from Cape Canaveral from three dozen this year to 50 or more in 2005. The fact that many of those missions will be carried out on new rockets is prompting state officials to scrutinize the Air Force's proposed change in safety procedures. The issue of protecting Floridians from launches dates to 1950, the year after Congress passed a law that led to the establishment of Cape Canaveral as a proving ground for guided missiles. Since then, the congressionally mandated protection job has been carried out by safety engineers at the local launch bases. The job entails making sure that missiles, rockets and piloted spaceships do not endanger launch workers, public and private property, or people living in Florida and in countries located beneath a rocket's flight path. If the Air Force plan takes effect, local safety officials no longer will have the power to demand design changes when flawed systems are discovered on rockets launching from Cape Canaveral. A second, but no less critical, part of the safety job is making certain that launch pads, satellite processing centers, hazardous fuel depots and ground support systems at Cape Canaveral are safely designed. Those responsibilities also would be transferred to Ohio. Others in the federal government are calling for the Air Force to stall the proposed move until the issue can be studied by the National Academy of Sciences. The academy is a nonprofit society of scholars that advises the U.S. government on science and technology issues. The review is expected to take nine months and cost $388,000. It will be paid for by the Air Force, NASA, two aerospace companies and the Federal Aviation Administration. ["Rocket safety plan criticized," **Florida Today**, January 13, 1999, p 1A & 2A.]

◆ A 72-page report commissioned by NASA and released Tuesday by the National Research Council gave the space agency high marks for the way it plans upgrades to the aging shuttle fleet. But the study was less enthusiastic about some of the specific improvements proposed, including a couple that have been under development for years. Two questions, experts suggested, should be considered before tackling any costly, future modifications: How likely is a replacement for the shuttle in the near future? What would the shuttle be used for after the projected end of the International Space Station era in 2013? "If they can develop a brand new vehicle that satisfies all of the requirements, they should do that," said Bryan O'Connor, an aerospace consultant and former astronaut who headed the 11-member committee. "If we wind up flying the shuttle out to 2030, but we've never really decided to do that, then the major cost savings and performance improvement upgrades will never be done. We'll kind of limp out to 2030." Minor shuttle modifications to improve safety and support flights continue to take place. But big-ticket renovations that would improve the shuttle's capabilities and made some basic changes have been mostly on hold since 1995. That's when a shuttle-management review headed by former Johnson Space Center Director Chris Kraft recommended freezing any shuttle design changes to save money. Since 1997, NASA has relied on about $100 million in reserve funds each year to pay for minimal upgrades. At the request of the Office of Management and Budget, NASA and five aerospace companies are putting together separate "space architecture" studies that are due out this spring. The studies will look at new spacecraft designs, how the programs would operate and the facilities needed to support them. Congress and the White House are expected to decide whether to replace the shuttle with a new, reusable vehicle by the end of the year 2000. ["What lies ahead for aging space shuttle?" **The Orlando Sentinel**, January 13, 1999, p A-6.]

JANUARY 13: In a hotly debated policy shift, the Air Force might scrap its practice of stopping rocket launches from Cape Canaveral until stray aircraft and boats are chased out of danger zones. Instead, the Air Force would continue its routine of broadcasting warnings to aviators and mariners in advance of a liftoff. But rather than risk costly launch postponements, unmanned rockets and shuttles would be allowed to fly despite the presence of ships and aircraft in restricted areas. Aerospace safety experts say the proposal could put private aircraft, fishing boats, shrimp trawlers and cruise ships at greater risk should they wander into zones that are off-limits during launches. Air Force officials are unwilling to speculate on possible public backlash against the plan. "I wouldn't dare guess what the public reaction would be," said Kevin Chilton, assistant director of operations at Air Force Space Command in Colorado Springs, Colo., which oversees operations at Cape Canaveral Air Station. "The question is if adequate warning is given and ignored, should you put at risk the launch vehicle – and the people in the vehicle for a manned flight – for the person who chooses to ignore the warning?" The possible change is part of a larger effort to cut the cost of doing business at Cape Canaveral Air Station, which has lost lucrative satellite launch contracts to foreign spaceports the past two decades. The issue is one of many that soon will be reviewed by the National Academy of Sciences as part of the first comprehensive study of the Air Force's range safety system, which has protected Floridians from exploding rockets for almost 50 years. ["Launches could put planes, boats at risk," **Florida Today**, January 14, 1999, p 1A & 2A.]

JANUARY 15: A damaged wire harness that escaped detection during preflight inspections triggered one of the costliest launch explosions in U.S. history, Air Force officials said Friday. A $344 million Titan 4 and its cargo, a spy satellite worth $700 million, erupted into a giant fireball 41 seconds into an Aug. 12, 1998 flight. The $1 billion blast showered flaming debris into a launch zone 3 miles off the coast of Cape Canaveral. In a report released Friday, investigators blamed the failure on electrical shorts within a wire harness designed to route power to the rocket's guidance system. Consequently, the flow of electricity to the guidance system was interrupted. That caused the rocket to pitch nose down, triggering its automatic destruct system. Investigators said there is "clear and convincing evidence" the harness was damaged before launch. Manufacturing records show inspectors found and fixed 44 other wiring defects on the harness before the flight. Flaws also were found on other identical Titan harnesses. Two Titan 4's are on Cape Canaveral launch pads. The first of those is tentatively scheduled for launch March 18. Air Force officials said neither of the rockets will fly before extensive analysis of harness flaws is completed. ["Faulty wire harness caused costly Titan explosion," Florida Today, January 16, 1999, p 1A.]

JANUARY 16: Over the next 12 months, six launches are scheduled using the NASA space shuttle. On April 8, Columbia is scheduled to take off with a crew of five astronauts and an X-ray observatory that will enable astronomers to study the universe. On May 13, Discovery and a crew of seven astronauts will take off on a mission to haul supplies up to NASA's International Space Station. Atlantis and five astronauts will be next up for NASA on Aug. 5. This will also be a space station construction mission. Then Endeavour is scheduled for Sept. 16. Four Americans and astronauts from both Europe and Japan will take off on their mission to make a radar map of the Earth. Spacewalking construction workers will head back to NASA's new international space station when Discovery and seven astronauts take off on Oct. 28. NASA will cap the year with the launch of shuttle Atlantis and five astronauts on Dec. 2 to complete another space station construction flight. ["Only 6 chances to watch shuttle launches in '99," Florida Today, January 17, 1999, p 6E.]

JANUARY 20: NASA's much-delayed launch of a $1.3 billion X-ray telescope will be pushed back at least another month so faulty circuit boards can be replaced, agency officials said Wednesday. Set for launch in April aboard shuttle Columbia, the trouble-prone Chandra X-Ray Observatory instead will remain on the ground until mid-May or later so repairs can be made. The problem – which could have crippled the telescope in orbit – is the latest in a series of glitches that have postponed its originally scheduled August 1998 launch from Kennedy Space Center. ["Telescope's flight delayed," Florida Today, January 21, 1999, p 1A & 2A.]

JANUARY 21: Space Shuttle Status Report, Thursday, January 21, 1999. STS-93: Chandra X-ray Observatory (formerly AXAF). KSC managers are reviewing Columbia's STS-93 processing schedule to accommodate a delay of at least one week in the Chandra payload KSC delivery date, previously slated for Jan. 28. NASA and Chandra spacecraft managers expect circuit board problems onboard Chandra to delay the STS-93 launch at least 5 weeks. Further spacecraft tests and inspections this week in Redondo Beach, California will allow Chandra managers to more accurately determine the schedule impact. That information will help NASA Shuttle managers address possible impacts to planned, downstream Shuttle flights. Yesterday, workers completed payload bay cleaning efforts and closed Columbia's payload bay doors. Orbiter forward and aft compartment closeouts continue and potable water servicing is under way. STS-96: 2nd U.S. International Space Station Flight Spacehab Double Module. Interface verification testing on Discovery's forward reaction control system is complete. Functional tests on orbiter hatch "D" have also concluded. Replacement of windows No. 2 and No. 7 continues. Heat shield modification efforts are under way and sensor work on the main propulsion system is ongoing. Preparations are in progress to replace 6 thrusters on Discovery's maneuvering system this week. STS-101: 3rd International Space Station Flight Spacehab Double Module / ICC. Atlantis remains in VAB high bay 2 awaiting Columbia's move out of OPF bay 3. installation of Atlantis' onboard tactical air command and navigation system (TACAN) is in progress and aft compartment leak checks are in work. Atlantis is scheduled to undergo orbiter processing for STS-101 in OPF bay 3. The date for Atlantis' transfer from the VAB to the OPF will be determined after Shuttle managers assess the impact of Shuttle Columbia's STS-93 delay. STS-99: Shuttle Radar Topography Mission (SRTM). Checks on Endeavour's forward reaction control system (FRCS) are complete. Workers are preparing to remove the FRCS this week. Heat shield and robot arm removal efforts are on going. Preparations for Shuttle main engine removal are also under way. Bruce Buckingham. (1999). **Kennedy Space Center Space Shuttle Status Report** [Online]. Available E-mail: domo@news.ksc.nasa.gov/subscribe shuttle-status [1999, January 21].]

JANUARY 22: Grady Williams, a former NASA design engineer died Friday. Williams worked on the first launch of an American ballistic missile, the Redstone, in August 1953 and later Apollo missions. He also worked on the start of the shuttle launches before he retired. Williams held a number of leadership positions in the space program, including director of KSC design and engineering from 1970 to 1972. ["KSC space pioneer Grady Williams dies," Florida Today, January 23, 1999, p 2B.]

JANUARY 24: A Lockheed Martin Athena 1 rocket is to carry Taiwan's first science satellite into orbit Tuesday night from a state-run launch pad at Cape Canaveral Air Station. The flight will be the second Athena mission from Florida. The first involved last year's launch of NASA's Lunar Prospector, which found evidence of frozen water on the moon. A third mission is planned for early 2000 when one of the vehicles is to carry an experimental military satellite into orbit from Cape Canaveral. ["Spaceport looks for boost in business," Florida Today, January 25, 1999, p 1A & 2A.]

JANUARY 25: Space Shuttle Status Report, Monday, January 25, 1999. STS-93: Chandra X-ray Observatory (formerly AXAF). Columbia's forward and aft compartment closeouts continue and potable water servicing is under way. The orbiter's drag chute compartment is undergoing instrumentation work this week. A system of temporary sensors is being installed inside the compartment to help engineers better understand the structural stress associated with vehicle transfers. This week, Shuttle managers are awaiting information from Chandra spacecraft officials to determine Columbia's target launch date and any impacts to downstream Shuttle flights. STS-96: 2nd U.S. International Space Station Flight Spacehab Double Module. Four of six orbital maneuvering system thrusters onboard Discovery were replaced over the weekend. The remaining two thrusters will be installed tonight. Replacement of orbiter windows No. 2 and No. 7 continues. Heat shield modification efforts are under way and sensor work on the main propulsion system is ongoing. This week, workers will test Discovery's power reactant storage and distribution system. STS-101: 3rd International Space Station Flight Spacehab Double Moldule / ICC. Atlantis remains in VAB high bay 2 awaiting Columbia's move out of OPF bay 3. Installation of Atlantis' onboard tactical air navigation system (TACAN) is in progress and aft compartment leak checks are in work. Atlantis is scheduled to undergo orbiter processing for STS-101 in OPF bay 3. The date for Atlantis' transfer from the VAB to the OPF will be determined after Shuttle managers assess the impact of Shuttle Columbia's STS-93 delay. STS-99: Shuttle Radar Topography Mission (SRTM). Endeavour's forward reaction control system will be removed from the orbiter this week. Robot arm removal is in work today and Shuttle main engine removal begins tonight. Docking mechanism removal is scheduled today. Bruce Buckingham. (1999). **Kennedy Space Center Space Shuttle Status Report** [Online]. Available E-mail: domo@news.ksc.nasa.gov/subscribe shuttle-status [1999, January 25].]

JANUARY 26: Launch of the spacecraft Stardust aboard a Boeing Delta II rocket is on track for Feb. 6 at 4:07 p.m. at the Cape Canaveral Air Station. The spacecraft is destined for a close encounter with the comet Wild 2 in January 2004. Using a silicon-based substance called aerogel, Stardust will capture comet particles flying off the nucleus of the comet. The spacecraft also will bring back samples of interstellar dust. The collected samples will return to Earth in a sample return capsule to be jettisoned as Stardust swings by Earth in January 2006. ["Stardust on track for launch on Feb. 6," **KSC Countdown**, January 26, 1999.]

◆ The southeastern beach mouse, an endangered species that has vanished from South Brevard County beaches, seems to be on the brink of dying out in Indian River County as well. That is the conclusion of Lew Erihardt, a biology professor at the University of Central Florida who annually conducts population studies of the tiny mammal. During a study last week, Erihardt found no beach mice south of Sebastian Inlet State Park. If there is no viable population of beach mice in South Brevard or Indian River counties, that leaves just a small number on government land at Kennedy Space Center, Canaveral National Seashore and Cape Canaveral Air Station. The mouse is doing well on government land because it is protected from development, said Donna Addy, a wildlife biologist for Dynamac Corp., which NASA contracts for environmental research. Addy said 175 mice were caught during a two-day population study in 1997. ["Beach mouse disappearing," **Florida Today**, January 26, 1999, p 1B.]

JANUARY 26: Florida could lose the biggest economic development battle in America, a key Tallahassee lawmaker wants to create a $30 million trust fund to help bring the next-generation spaceship to the state. The project in question is Lockheed Martin's VentureStar, a spaceship expected to cut the high cost of launching satellites and astronauts into orbit as much as 90 percent. In doing so, the new ship could render obsolete the expensive unmanned rockets and piloted shuttles now launched from Cape Canaveral Air Station and Kennedy Space Center. That, in turn, could shut Florida out of a highly competitive global business that could generate $20 billion a year by 2010. With a $941 million NASA contract in hand, and $200 million of its own money to invest, Lockheed Martin is building a half-scale VentureStar prototype. A spate of technical problems has delayed until next year test flights of the X-33, which will take place in California. Nonetheless, Lockheed Martin officials say they plan to begin launching the full-scale VentureStar in late 2004. California and Florida are thought to be front-runners in the nationwide bidding war to become VentureStar's launch site. Sixteen other states also are interested, including New Mexico, Montana, Virginia and Oklahoma. Delays in the X-33 test flight program are expected to push back Lockheed Martin's launch site decision from late this year to March 2000. ["Lawmaker proposes $30 million trust fund to lure space projects," Florida Today, January 27, 1999, p 10C & 9C.]

◆ Charles Luckman, an entrepreneur and architect who designed skyscrapers and parts of Kennedy Space Center, among other landmarks, has died. He was 89. In addition to KSC, he also created Johnson Space Center in Houston, the new Madison Square Garden in New York and the Prudential Center in Boston. ["KSC architect dead at 89," Florida Today, January 27, 1999, p 3A.]

◆ A Lockheed Martin rocket successfully lifted off Tuesday night from Cape Canaveral Air Station, carrying Taiwan's first science satellite into orbit. The rocket left the pad at 7:34 p.m., marking just the second time an Athena has flown from Florida. It was also the second launch from Spaceport Florida Authority's new launch complex at the air station. ["Athena rocket sends satellite into night sky," Florida Today, January 27, 1999, p 2A.]

◆ Kennedy Space Center soon will host a major space film production. *Daily Variety* says Clint Eastwood will direct and co-star with Tommy Lee Jones in a Warner Bros. action space film this summer. The film, called *Space Cowboys*, is about two older guys drafted by NASA to fly a crucial shuttle mission decades after their crack piloting days, when they didn't make the astronaut program because they were too cocky. ["Space Center may house Space Cowboys," The Orlando Sentinel, January 27, 1999, p A-2.]

JANUARY 28: Gov. Jeb Bush said Thursday that no project is more important to Florida's economic development than efforts to lure a next-generation spaceship called VentureStart to the state. "The most important (recruitment project) on the horizon today is VentureStar," Bush told members of Enterprise Florida at a Tallahassee economic summit. "We need to respond to these... economic opportunities." Enterprise Florida is coordinating the move to try to bring VentureStar to Florida, where it would take off and land at NASA's Kennedy Space Center or Cape Canaveral

Air Station. ["Bush: Luring VentureStar top priority," <u>Florida Today</u>, January 29, 1999, p 1A & 2A.]

◆ Spaceport Florida Authority announced Thursday that it will build a rocket assembly building to be leased to the Boeing Co. The state-run authority will construct the $15 million, 102,000-square-foot Horizontal Integration Facility. Boeing will equip it for the assembly and testing of its Delta 4 rockets. The Cape Canaveral Air Station facility will be adjacent to Launch Complex 37, which is undergoing major renovation for that purpose. Complex 37 will be one of two complexes used to launch the new rocket. The other is at Vandenberg Air Force Base in California. Boeing expects to begin using the building in July 2000, with the first launch of a Delta 4 expected some time in 2001. The agreement Boeing has with the authority calls for a 20-year lease to cover the construction of the building as well as related administrative costs. Parties to the deal expect the lease to be signed March 1. ["Boeing will use new rocket assembly plant," <u>Florida Today</u>, January 29, 1999, p 1C.]

JANUARY 31: Longtime McDonnell Douglas executive, George Faenza, 67, died Sunday in an Atlanta hospital from a heart attack. Before retiring in 1996, he was McDonnell Douglas' vice president and general manager at Kennedy Space Center. ["Space pioneer, child advocate George Faenza dies," <u>Florida Today</u>, February 2, 1999, p 1B.]

FEBRUARY

FEBRUARY 1: Space Shuttle Status Report, Monday, February 1, 1999. STS-93: Friday, workers completed efforts to pressurize Columbia's reaction control system manifold. A system of temporary sensors have been installed inside the orbiter's drag chute compartment to help engineers better understand the structural stress the vehicle is exposed to preflight. The Chandra payload remains scheduled to arrive at KSC Thursday evening. Orbiter rollover to the Vehicle Assembly Building is currently under review. STS-96: Discovery's orbital maneuvering system thruster replacement is complete and workers are now replacing the thruster bellows. Orbiter window replacement efforts continue. Heat shield modification efforts and sensor work on the main propulsion system continue. STS-101: Atlantis remains in VAB high bay 2 awaiting Columbia's move out of OPF bay 3. Installation of Atlantis' onboard tactical air and navigation system (TACAN) is in progress and aft compartment leak checks are in work. Atlantis is scheduled to undergo orbiter processing for STS-101 in OPF bay 3. The date for Atlantis' transfer from the VAB to the OPF will be determined after Shuttle managers assess the impact of Shuttle Columbia's STS-93 delay. STS-99: Endeavour's main engines were removed last week. Forward reaction control system removal also concluded last week. Preparation for tomorrow's removal of the external airlock is under way. Verification of the robot arm's pyrotechnic initiator controllers is scheduled this week. Bruce Buckingham. (1999). **Kennedy Space Center Space Shuttle Status Report** [Online]. Available E-mail: domo@news.ksc.nasa.gov/subscribe shuttle-status [1999, February 1].]

◆ The 45th Space Wing at Patrick Air Force Base was honored Monday with the Air Force Outstanding Unit Award at a ceremony at the base. ["45th Space Wing honored for its work on launches," <u>Florida Today</u>, February 2, 1999, p 2B.]

◆ The Clinton administration wants to spend an extra $2 billion in the next five years to keep the International Space Station from falling victim to Russia's economic crisis, NASA officials said Monday. More than half the money would come by cutting other agency programs and using it to pay for station equipment the Russians had planned to build but now cannot afford. For the remainder, the White House has agreed to commit $800 million through 2005 for a "Russian contingency fund." Neither NASA officials nor President Clinton's fiscal year 2000 budget proposal, which was released Monday, identified the source of the extra $800 million. In all, Clinton wants to give NASA $13.57 billion, including $3 billion for the space shuttle program and $2.5 billion for the station. The money is a small part of Clinton's $1.77 trillion federal budget. ["Clinton: Save space station," <u>Florida Today</u>, February 2, 1999, p 1A.]

FEBRUARY 2: Lockheed Martin won't make a third attempt tonight to launch an Atlas rocket and Japanese communications satellite from Cape Canaveral. Monday's attempt was postponed so workers could make a precautionary inspection of parts in the rocket's engine system. The launch had been postponed earlier from Sunday night. The inspection was ordered after similar parts in the factory were found to be incorrectly assembled. Company officials haven't decided when to try again. ["Atlas launch on hold for now," <u>Florida Today</u>, February 2, 1999, p 1A.]

◆ The launch of NASA's Stardust spacecraft aboard a Boeing Delta II rocket is scheduled for Saturday, Feb. 6. There is a single instantaneous launch opportunity available that day at 4:06:42 p.m. The next available window is on Sunday, Feb. 7 at 4:04:15 p.m. Liftoff will occur from Launch pad 17-A, Cape Canaveral Air Station. Stardust has completed final checkout at KSC and was mated to the Boeing Delta II at the launch pad on Jan. 28. The Delta fairing is being installed around the spacecraft on Feb. 2. ["Stardust to launch on Saturday for rendezvous with comet in 2004," **KSC Countdown**, February 2, 1999.]

◆ In his first budget proposal, Governor Jeb Bush has proposed spending at least $10 million to help attract VentureStar, a next-generation spaceship, to Florida. Last year, Florida allocated $1.6 million for the project. ["Bush seeks more money, hefty tax cut," <u>Florida Today</u>, February 3, 1999, p 1A & 2A.]

◆ The Air Force said Tuesday it has figured out how to protect its Titan rockets from the kind of catastrophic accident that caused a $1 billion launch failure from Cape Canaveral in August. In releasing the list of "corrective actions," the Air Force said it would be ready to launch a Titan again in March from Cape Canaveral Air Station. Investigators determined that electrical shorts in the rocket's power supply wiring harness most likely caused the accident. The plan for corrective actions includes reinspecting all wire harnesses on Titan launch vehicles, and redesigning or modifying systems related to vehicle power and guidance, the Air Force said. ["Titan rockets nearly ready for service, Air Force says," <u>Florida Today</u>, February 3, 1999, p 2A.]

FEBRUARY 3: Lisa Malone, NASA's media chief at Kennedy Space Center, will be working at Melbourne-based public relations and marketing firm Watermark Communications through April. It's part of a NASA training program designed to broaden a worker's experience through course work and temporary job assignments. Called Senior Executive Service Candidate Development, the program is meant to groom NASA employees for high-level positions. Chief of KSC's newsroom, Malone started the training program last year with a stint in the White House's office of science and technology policy. The training program will continue through the year, giving Malone the chance to take a number of classes and job assignments. ["Media chief gets training," <u>Florida Today</u>, February 4, 1999, p 12C.]

FEBRUARY 4: Cutbacks and a long hiring freeze have left NASA's work force so thin the agency is confronting a "brain drain" that could jeopardize the safety of space shuttle flights, an independent board reported Thursday. Technically, NASA centers have been given authority to resume hiring. But budget ceilings are making it all but impossible to fill jobs. As a result, Kennedy Space Center, Johnson Space Center in Houston and Marshall Space Flight Center in Huntsville, Ala., cannot sign fresh talent until at least 2001, the Aerospace Safety Advisory Panel said. KSC, Johnson and Marshall, which together run the shuttle program, each face additional losses of 300 to 400 positions beyond the significant downsizing they have experienced already. Another worry, particularly at KSC, is NASA's supervisory work force will lose skills needed to ensure safe shuttle operations as the United Space Alliance takes over responsibility for daily operation of the fleet. Goldin acknowledged the effect continued downsizing is having, but told those attending the meeting that safety remains NASA's top priority. ["Thin work force could threaten shuttle safety, panel reports," **Florida Today**, February 5, 1999, p 1A.]

◆ A $.15 billion telescope arrived from California Thursday at Kennedy Space Center, facing repairs on its circuit boards before it can be launched this summer aboard shuttle Columbia. The Chandra X-Ray Observatory, which was scheduled to fly in April, will remain on the ground indefinitely so the faulty parts can be replaced. NASA has not set a new launch date, but agency documents show the telescope could be carried into orbit in early July. ["New telescope arrives at KSC, faces repairs," **Florida Today**, February 5, 1999, p 5A.]

FEBRUARY 5: Last year, for the first time this decade, NASA's Kennedy Space Center pumped less than $1 billion into Florida's economy, a victim of federal budget cuts and massive restructuring. In its annual report released Friday, NASA said the space center funneled $966 million into Florida in the year ended Sept. 30, 4 percent less than the previous fiscal year. This is the fifth straight year KSC's economic impact has declined. In fiscal 1993, the space center had an economic impact of $1.5 billion; that dropped to 1.03 billion in fiscal 1997. "As the budget has gone down for NASA... KSC has also seen budget cuts," said Joel Wells, NASA's spokesman at KSC. He said the lower budget has a direct effect on what KSC can spend in the state and in Brevard County. However, Brevard County is getting a larger share of the shrinking pie. About 83 percent of 1998's spending -- $799 million - went to Brevard, $17.5 million more than last year. Most of that money -- $762 million - went to companies operating at the space center. An additional $37 million went to off-site businesses in Brevard. At the end of 1998, KSC employed 11,984 workers. In 1993, it workforce totaled 18,253. ["Space center added less to state coffers," **Florida Today**, February 6, 1999, p 12C.]

◆ Russia's deeply troubled partnership in the International Space Station is forcing NASA to reduce the number of shuttle flights this year from six to five, the agency announced Friday. Consequently, the station's first full-time crew won't climb aboard the outpost until February 2000, a month later than planned. The changes are part of a new flight schedule triggered by another delay in the launch of a crucial Russian command and control module to the station, the first two parts of which lifted off late last year. Several shuttle missions devoted to station assembly work cannot fly until the Russian module reaches orbit. The latest plan is for NASA to flip-flop the first two shuttle flights this year because of the late delivery of an observatory that will be used to study stars and galaxies. NASA's first flight is scheduled for May 20, when a crew aboard Discovery ferries supplies to the new station. Facing a longer delay is a mission to launch the $1.3 billion Chandra X-Ray Observatory on shuttle Columbia. Once scheduled for launch in August, that flight is set for July 9. Next is a Sept. 16 shuttle Endeavour mission to make a radar map of the Earth. Already more than a year behind schedule, the key Russian piece won't be flown from Russian until Sept. 20 or later. Among other things, it will provide living quarters for the station's first regular crew. NASA will follow up with station construction missions aboard Atlantis on Oct. 14 and Discovery on Dec. 2. Another shuttle station flight originally set for this year has been postponed to Feb. 3, 2000. ["NASA cuts '99 shuttle flights to 5," <u>Florida Today</u>, February 6, 1999, p 1A.]

FEBRUARY 6: A glitzy gateway modeled after the International Space Station. Robots that crack jokes. A movie showing NASA's dramatic quest to find life on other planets with big-screen sights and sounds. This is Kennedy Space Center, where tourism humbly began in 1963 when Cape Canaveral was open for three hours on Sunday for motorists who used booklets for self-guided tours. In April, four new attractions from a talking robot feature to a movie about the search for life in space will open at Kennedy Space Center Visitor Complex. Two more attractions are set to open in the fall. The projects will top off a $100 million makeover that started in 1995. More improvements will be conceived under a 10-year redevelopment plan that is expected to get under way by the end of the year. Kennedy Space Center Visitor Complex attendance: 1998 -- 2.75 million; 1997 -- 2.70 million; 1996 -- 2.40 million; 1995 -- 2.15 million; 1994 -- 2.05 million; 1993 -- 2.45 million; 1992 -- 2.76 million; 1991 -- 2.63 million; 1990 -- 3.10 million. ["KSC expands Visitor Complex," <u>Florida Today</u>, February 7, 1999, p 1E & 5E.]

◆ Technical problems postponed the launch of NASA's Stardust probe from Cape Canaveral Air Station on Saturday. Launch controllers were uncertain whether a radar beacon on the Delta that ground stations use to track the rocket's flight path was working properly. The countdown was stopped with less than two minutes remaining. Mission managers tentatively plan to make a second try during a split-second launch window at 4:04:15 p.m. today. Once in space, Stardust will begin a 2.3 billion-mile voyage to rendezvous with Comet Wild-2. The probe will attempt to capture dust particles from the comet in 2004 and return them to Earth in 2006. ["Technical problems delay launch of Stardust probe," <u>The Orlando Sentinel</u>, February 7, 1999, p A-3.]

◆ The Chandra X-ray Observatory was unloaded on Saturday, Feb. 6, and transported to the Vertical Processing Facility (VPF). It arrived aboard an Air Force C-5 Galaxy aircraft at the Shuttle Landing Facility on Thursday, Feb. 4. At the VPF, the telescope will undergo final installation of associated electronic components; it will also be tested, fueled and mated with the Inertial Upper Stage booster. A set of integrated tests will follow. ["Chandra X-ray Observatory unloaded, moved to Vertical Processing Facility," **KSC Countdown**, February 9, 1999.]

FEBRUARY 7: A Delta 2 rocket lifted off from Cape Canaveral Air Station on Sunday afternoon carrying the $128 million Stardust spacecraft. ["Rendezvous with destiny," <u>The Orlando Sentinel</u>, February 8, 1999, p A-1.]

FEBRUARY 8: Space Shuttle Status Report, Monday, February 8, 1999. STS-96: Following Discovery's recent orbital maneuvering system thruster replacements, workers completed manifold draining and backfilling efforts over the weekend. Troubleshooting of a minor leak on APU No. 2 is under way. Installation of the androgynous peripheral docking system (APDS) begins today. This week, technicians will resume efforts to install the integrated vehicle health monitoring system on the orbiter's main propulsion system. STS-93: Technicians have completed Columbia's processing efforts to date and are preparing to transport the orbiter to the Vehicle Assembly Building on Wednesday. Columbia will reside in VAB high bay 2 in temporary storage until mid-April, when Shuttle Discovery rolls out of OPF bay 1. Columbia will then be transferred to OPF bay 1 to complete STS-93 orbiter preparations. STS-99: Removal of Endeavour's tunnel adapter forward extension is complete. Water spray boiler No. 3 installation is also complete. Workers are servicing freon coolant loop No. 1 today and efforts are under way to upgrade the main propulsion system's 17-inch disconnect valve actuator. Tunnel adapter removal is also scheduled this week. STS-101: Atlantis is being moved to the VAB transfer aisle today to allow room for Shuttle Columbia's shift into VAB high bay 2. Atlantis will remain in the transfer aisle while preparations are made for its arrival in OPF bay 3 on Feb. 17. A majority of the orbiter's tactical air and navigation system (TACAN) modification is complete. Only 2 weeks of TACAN work remains. Workers also completed about half of the required orbiter leak checks while in the VAB. Bruce Buckingham. (1999). **Kennedy Space Center Space Shuttle Status Report** [Online]. Available E-mail: domo@news.ksc.nasa.gov/subscribe shuttle-status [1999, February 8].]

◆ NASA's Cassini spacecraft is on course to fly near Earth this summer, and opponents of the plutonium-powered journey already are gearing up to protest. Cassini's $3.2 billion mission to Saturn will be the focus of a modest Feb. 27 rally planned by two anti-nuclear groups outside the gates of Cape Canaveral Air Station. Organizers expect no more than 20 people to attend the rally but believe it will send a message. Launched from Cape Canaveral in October 1997, Cassini is flying a roundabout course that will take it to Saturn in July 2004. Before it can get to Saturn, Cassini must fly near several planets and use their gravity to increase its speed. The spacecraft made a pass by Venus last year; a second Venus approach is set for June. The Earth flyby will follow Aug. 18. A final maneuver near Jupiter is set for

December 2000. NASA engineers say there is a less than 1-in-1 million chance the spacecraft will veer off course and crash back through Earth's atmosphere. ["Cassini opponents plan Feb. 27 protest rally," <u>Florida Today</u>, February 9, 1999, p 1B.]

FEBRUARY 10: The state and NASA are banding together to build a world-class test lab that could help lure next-generation spaceships to Florida and lead to rocket fuel factories on Mars. State officials gave $750,000 on Wednesday to construct the lab, which will specialize in developing advanced systems for spaceships that use cryogenic – or supercold – liquid fuel to reach orbit. NASA officials added $1.5 million that will be use to outfit the lab with state-of-the-art test equipment. The joint effort could give Florida an edge in a nationwide bidding battle for Lockheed Martin's VentureStar, a new liquid-fueled launcher that could replace the rockets and shuttles that now fly from KSC and Cape Canaveral Air Station. "That's exactly our thinking," said Karen Thompson, NASA's manager of technology testbeds at KSC. "That was one of our selling points." Dubbed the Cryogenics Testbed, the lab will provide government, industry and commercial customers with a place to develop advanced systems for use with supercold propellants such as liquid hydrogen and liquid oxygen. NASA's space shuttles and other rockets use the propellants. So will VentureStar, and officials think the KSC lab might help sway Lockheed Martin when they make a decision next year on where to base the ship. The lab will be operated by Dynacs Engineering Co. Inc., NASA's engineering support contractor at KSC. NASA plans to use the lab to help develop the technologies that will be needed to build rocket fuel and water-generation plants on the surface of Mars. Both efforts are deemed key to preparing for human expeditions to the Red planet. Other partners in the project include the University of Florida and Air Products and Chemicals Inc., NASA's prime supplier of liquid hydrogen for the shuttle program. The lab should be open in a year. ["NASA, state join forces to build high-tech test lab," <u>Florida Today</u>, February 11, 1999, p 2A.]

FEBRUARY 11: Space Shuttle Status Report, Thursday, February 11, 1999. STS-96: Technicians have completed checks of Discovery's recently replaced orbital maneuvering system (OMS) thrusters. Workers have completed preparations to install the orbiter's new docking system and installation efforts are ongoing. Installation of the integrated vehicle health monitoring system for the orbiter's main propulsion system is in progress and base heat shield modifications continue. STS-93: Yesterday, Columbia rolled out of OPF bay 3 and into VAB high bay 2 for temporary storage. Shuttle Atlantis will be the next occupant in OPF bay 3, following a week of open bay preparation. Columbia will remain in the VAB until mid-April, when Shuttle Discovery rolls out of OPF bay 1. Columbia will then be transferred to OPF bay 1 to complete STS-93's orbiter preparations. STS-99: Removal of the SAC-A payload from Endeavour's cargo bay is complete. Deservicing of the orbiter's freon coolant loop No. 1 is complete and valve modification is now in work. Technicians are conducting post-installation inspections on water spray boiler No. 3. Efforts to upgrade the main propulsion system's 17-inch disconnect valve actuator continue. STS-101: Atlantis remains in the north end of the VAB transfer aisle. The orbiter waits in the transfer aisle while preparations are made for its arrival in OPF bay 3 on Feb. 17. A majority of the orbiter's tactical air and navigation system (TACAN) modification is complete. Only two weeks of TACAN work remains. Workers also completed about half of the required orbiter leak checks while in the VAB. Bruce Buckingham. (1999). **Kennedy Space Center Space Shuttle Status Report** [Online]. Available E-mail: domo@news.ksc.nasa.gov/subscribe shuttle-status [1999, February 11].]

FEBRUARY 12: On Feb. 12, NASA announced a change in the crews on upcoming Space Shuttle missions STS-96 and STS-101. Russian cosmonaut Valery Tokarev has been named to the STS-96 mission, replacing commander Yuri Malenchenko, who is now assigned to STS-101. Also assigned to STS-101 is cosmonaut Boris Morukov. Tokarev joined other STS-96 crew members at KSC last week for a payload Interface Verification Test, checking equipment that will fly in the Spacehab logistics double module on the first logistics flight to the International Space Station. STS-96 is scheduled to launch on May 20. ["Crew changes announced for upcoming missions · STS-96, STS-101," **KSC Countdown**, February 18, 1999.]

FEBRUARY 14: Facing a shortage of parts and money, NASA is taking back part of an exhibit on the space shuttle from an Alabama museum, according to a newspaper report. The Marshall Space Flight Center and the United Space Alliance, NASA's shuttle operations contractor, last week contacted the U.S. Space & Rocket Center in Huntsville, Ala., and asked it to return the forward assemblies from the solid rocket boosters on the museum's full-size shuttle exhibit "for use in the space program," the *Huntsville Times* newspaper reported Sunday. The shuttle's solid rocket boosters serve as propellant motors for the spacecraft during launch. After launch, the separate from the shuttle at 200,000 feet and fall into the ocean, where they are recovered, towed back to shore and refurbished for another flight. But several of the forward assemblies used in the shuttle program have been damaged or lost since 1981, when the shuttle program began. NASA told the newspaper that it asked for the parts back to make sure the space agency does not fall short of hardware over the coming months as it constructs the new international space station. Officials at NASA were not available to comment Sunday. ["NASA asks museum to return hardware," **The Orlando Sentinel**, February 15, 1999, p A-3.]

FEBRUARY 15: A Lockheed Martin Atlas rocket successfully carried out a $200 million satellite delivery mission Monday night, ferrying a Japanese communications spacecraft into orbit. The rocket blasted off at 8:45 p.m. from Cape Canaveral Air Station, and deployed a satellite called JCSAT-6 about half an hour later. The successful liftoff ended a string of delays for the mission, which had been scheduled for launch Jan. 31. The liftoff was postponed two weeks while technicians investigated a possible problem with the motors that control the flow of fuel into the rocket's Centaur stage. Another delay was caused Sunday when controllers scrubbed the flight to check a possible fuel tank problem. The next launch from Cape Canaveral isn't until April 9 when an Air Force Titan rocket is scheduled to carry a military satellite into orbit to spot missile launches. ["Atlas delivers satellite to orbit," **Florida Today**, February 16, 1999, p 1A.]

◆ The U.S. Fish and Wildlife Service is accepting bids to trap and catch wild hogs on the Merritt Island National Wildlife Refuge. Officials will award three new contracts when the current contracts expire March 31. The application deadline is March 12. ["Refuge accepts bids for hog trappers," Florida Today, February 16, 1999, p 1B.]

FEBRUARY 16: In a move that could alter the way launch business is done around the world, the state next month will begin building a "quick response" launch complex at Cape Canaveral Air Station. The goal – get rockets assembled and off the ground in the unheard of time of six hours or less. Known as Launch Complex 20, the oceanside site is made up of a launch control blockhouse, the rusting remains of an old Titan test stand, a payload support building and two small, flat concrete launch pads. The hub of the renovated complex will be a unique drive-through launch center designed for small rockets mostly used to loft scientific experiments and instruments on short, suborbital jaunts. Launch companies will be able to haul rockets into the center, ready them for flight, and rush them out for launch. The project, which is scheduled to start March 1, should be completed by Sept. 1. It is expected to be open for business a month later. ["'Quick' launch site taking shape at Cape," Florida Today, February 17, 1999, p 1A & 2A.]

FEBRUARY 17: The Air Force should ditch its plan to modernize the way it tracks Florida rocket launches and let private industry do it using advanced 21st-century technology, U.S. Rep. Dave Weldon said Wednesday. The issue involves the Eastern Range, a vast Air Force-run tracking system that monitors all missions that blast off from Cape Canaveral Air Station and Kennedy Space Center. Experts who operate the system relay flight data to ground controllers. They also are responsible for destroying vehicles if they go astray and threaten populated areas. However, the system is stumbling along on 1950s and '60s technology that makes it hard for Florida to meet a growing demand by private companies to launch communications satellites. Although the Air Force is trying to upgrade the equipment, the plan – which is expected to cost about $1 billion on the East Coast - falls short, Weldon said. The congressman said there is an opportunity for private industry to take the lead later this year. That is when the contract for operating the range, held by Computer Sciences Raytheon, goes out for bid. Weldon said he will send a letter this week to Acting Air Force Secretary F. Whitten Petters, asking him to combine that annual $95 million contract with the contract in place to improve the range. In competing to win both contracts, companies could investigate new ways to make the entire operation more advanced, Weldon said. ["Weldon: Industry should update rocket range," Florida Today, February 18, 1999, p 1A.]

FEBRUARY 22: Space Shuttle Status Report, Monday, February 22, 1999. STS-96: Last week, technicians installed Discovery's remote manipulator system or robot arm in the payload bay. Over the weekend, workers began efforts to install additional sensors inside the orbiter's drag chute compartment. The added instrumentation will gather data through the first few seconds of liftoff. Installation of the integrated vehicle health monitoring system for the orbiter's main propulsion system continues and base heat shield modifications are on schedule. Payload premate test preparations are in work. Booster stacking operations are under way in VAB high bay 3. STS-93: Columbia is jacked and leveled in VAB high bay 2 undergoing routine system observation during a temporary storage period. Columbia will remain in the VAB until mid-April, when Shuttle Discovery rolls out of OPF bay 1. Columbia will then be transferred to OPF bay 1 to complete STS-93's orbiter pre-launch preparations. STS-99: Replacement of Endeavour's left hand radiator No. 1 is complete and radiator No. 2 is being replaced today. This work is part of a valve modification being conducted on freon coolant loop No. 1. STS-101: Atlantis moved from the VAB transfer aisle to OPF bay 3 on Feb. 17. Workers are setting up platforms to gain access to the orbiter. Once access is established, technicians will resume the remaining work on the tactical air and navigation system (TACAN). The orbiter's payload bay doors will be opened this week. Bruce Buckingham. (1999). **Kennedy Space Center Space Shuttle Status Report** [Online]. Available E-mail: domo@news.ksc.nasa.gov/subscribe shuttle-status [1999, February 22].]

◆ NASA might mount an emergency mission to repair the Hubble Space Telescope this fall to avoid a possible shutdown of the $3 billion observatory and its delivery of stunning discoveries. The reason: A system crucial to pointing the telescope is operating on its last legs, and its failure would shut down Hubble until a scheduled June 2000 servicing mission could be carried out. Senior NASA officials, as a result are thinking seriously about sending a spacewalking telescope repair crew on a shuttle rescue mission in October. The idea is to fix Hubble before it breaks, so scientists are not cut off from the flagship of orbital astronomy. A decision could be made within a week. ["Shuttle crew may be sent to fix Hubble in October," **Florida Today**, February 25, 1999, p 1A & 2A.]

FEBRUARY 25: KSC Chandra X-Ray Telescope Status Report, Thursday, February 25, 1999. In the Vertical Processing Facility (VPF) at KSC, work continues on schedule toward the launch of the Chandra X-ray telescope aboard Space Shuttle Columbia on July 9 on mission STS-93. The component rework of the Command and Telemetry Unit (CTU) is being completed this week. The CTU will arrive from the vendor Saturday, Feb. 27 and be re-installed into Chandra on Monday, March 1. The Interface Unit (IU) was removed from Chandra on Feb. 14 for rework and will be returned to KSC and re-installed on March 7. The CTU and IU will undergo testing March 8-12. A full state-of-health test for the telescope is scheduled to occur March 18. The attitude control thrusters will be helium flow-tested the following day. Planning is underway for the arrival of the solar arrays. They are being transported by C-17 aircraft to the Shuttle Landing Facility with arrival currently scheduled for March 9. Operations to mate the arrays to Chandra are scheduled for the week of March 23 and they will be fully deployed for testing on March 27. On March 31, Chandra will be hoisted from its test stand onto a fueling stand. Loading of fuel into the spacecraft is currently planned to begin on April 5 and will take ten days to complete. These bi-propellant hydrazine and nitrogen tetroxide fuels will be used by the telescope to achieve its final orbit. Hydrazine will also be used by one of the telescope's subsystems associated with pointing of the telescope. Finally, the spacecraft's batteries will be installed on April 17. Chandra will then be ready for the arrival at the VPF of the Inertial Upper Stage booster on Apr. 19. Chandra arrived at KSC's Shuttle Landing Facility aboard an Air Force C-5 airplane on Feb. 4. It was offloaded and transported to the VPF on Feb. 6. The telescope was removed from the shipping container on Feb. 8. It was rotated to the vertical position and placed in a test stand on Feb. 10. Bruce Buckingham. (1999). **KSC Chandra X-Ray Telescope Status Report** [Online]. Available E-mail: domo@news.ksc.nasa.gov/subscribe shuttle-status [1999, February 25].]

◆ Space Shuttle Status Report, Thursday, February 25, 1999. STS-96: Technicians will test Discovery's remote manipulator system or robot arm today. Replacement of the orbiter's water spray boiler is under way with removal of the old system already complete. Functional tests of the International Space Station docking system are scheduled this week. Installation of the integrated vehicle health monitoring system for the orbiter's main propulsion system continues and base heat shield modifications are on schedule. Drag chute instrumentation installation continues. Tests of Discovery's power reactant storage and distribution system and auxiliary power units are ongoing. Booster stacking operations continue in VAB high bay 3. STS-93: Columbia is jacked and leveled in VAB high bay 2 undergoing routine system observation during a temporary storage period. Columbia will remain in the VAB until mid-April, when Shuttle Discovery rolls out of OPF bay 1. Columbia will then be transferred to OPF bay 1 to complete STS-93's orbiter pre-launch preparations. STS-99: Modifications of Endeavour's freon coolant loop No. 1 continue to go well, with replacement of the left hand radiators now complete. The orbiter's forward reaction control system will arrive in the OPF today for installation inside the orbiter's nose on Saturday. STS-101: Atlantis' payload bay doors were opened yesterday and access to the orbiter's flight deck has been established. Work on the tactical air and navigation system (TACAN) has resumed. Atlantis' orbiter maneuvering and reaction control systems are undergoing drain line modifications. Bruce Buckingham. (1999). **Kennedy Space Center Space Shuttle Status Report** [Online]. Available E-mail: domo@news.ksc.nasa.gov/subscribe shuttle-status [1999, February 25].]

◆ A senior NASA official admitted Thursday that a key Russian piece for the International Space Station might not be launched until late this year, causing more trouble for the project. The 21-ton Russian service module is scheduled for liftoff no sooner than September but might fly as late as November or December, said Joe Rothenberg, NASA's associate administrator for the office of spaceflight. ["Russia may keep NASA waiting again," **Florida Today**, February 26, 1999, p 1A & 2A.]

FEBRUARY 27: Opponents of NASA's plutonium-powered Cassini spacecraft gathered Saturday outside the gates of the Cape Canaveral Air Station as they gained momentum for other protests against the use of nuclear power in space. About 20 protesters carrying signs rallied for about a half-hour under the watchful eyes of security officers with the Air Station, where Cassini was launched in 1997. ["Cassini opponents protect craft's August Earth flyby," **Florida Today**, February 28, 1999, p 1B.]

◆ Homestead Air Force Base is one of at least three Florida military bases being studied by the aerospace industry as an alternative to Cape Canaveral for launching satellites. Two others are in North Florida: Eglin Air Force Base in the state's Panhandle and Cecil Field Naval Air Station in Jacksonville. So far, the idea is little more than talk. Some of the proposals appear farfetched. All face enormous obstacles. As technology reshapes the booming space industry, Florida will lose business if it doesn't develop more commercial launch sites. Not long ago practically all of the Western world's launches were from Cape Canaveral Air Station and Kennedy Space Center. Today, competition comes from Russia, China and a European consortium in South America, less than a third of the world's satellites blast off from Florida's Space Coast. One industry analyst predicts that 1,700 payloads worth more than $120 billion will need a lift into orbit during the next decade, an average of 170 per year. Including manned flights, there were 81 launches worldwide in 1998. Only 22 were from the Cape. ["Another launch site for Florida?" **The Orlando Sentinel**, February 28, 1999, p A-1 & A-12.]

MARCH

MARCH 1: As of March 1, 1999, Lewis Research Center will be renamed John Glenn Research Center at Lewis Field. ["Lewis Research Center renamed to honor John Glenn," **KSC Countdown**, February 23, 1999.]

◆ Space Shuttle Status Report, Monday, March 1, 1999. STS-96: Last Friday, technicians completed verification tests on Discovery's remote manipulator system or robot arm. Today, workers will perform functional checks on the orbiter's new docking mechanism. Replacement of the orbiter's water spray boiler continues. Installation of the integrated vehicle health monitoring system for the orbiter's main propulsion system proceeds on schedule and base heat shield fastener modifications continue. Drag chute instrumentation installation continues as well. Booster stacking operations continue in VAB high bay 3. STS-93: Columbia is jacked and leveled in VAB high bay 2 undergoing system observation during a temporary storage period. Columbia will remain in the VAB until mid-April, when Shuttle Discovery rolls out of OPF bay 1. Columbia will then be transferred to OPF bay 1 to complete STS-93's orbiter pre-launch preparations. STS-99: Shuttle Endeavour's forward reaction control system has been installed and preparations for functional checks are ongoing. Modifications of Endeavour's freon coolant loop No. 1 continue to go well. STS-101: Work on Atlantis' tactical air and navigation system (TACAN) is under way. Atlantis' orbiter maneuvering and reaction control systems are undergoing drain line modifications and window No. 1 is being replaced. Bruce Buckingham. (1999). **Kennedy Space Center Space Shuttle Status Report** [Online]. Available E-mail: domo@news.ksc.nasa.gov/subscribe shuttle-status [1999, March 1].]

◆ An investigation into one of the costliest launch failures in U.S. space program history has prompted the Air Force to withhold tens of millions of dollars from rocket maker Lockheed Martin. The action was prompted by investigators who have uncovered serious quality control problems and questionable work practices throughout the Lockheed-run Titan rocket program, which is crucial to national security. As a result, the company made no profit for work done from July through December 1998 on two contracts it has to build and launch the vehicles, according to Air Force documents obtained by *Florida Today*. The penalty followed the Aug. 12, 1998, launch explosion at Cape Canaveral Air Station that destroyed a $344 million Titan 4 and a top-secret National Reconnaissance Office spy satellite worth $700 million. ["Air Force won't pay millions to Lockheed," **Florida Today**, March 2, 1999, 1A & 2A.]

◆ The prototype of a reusable, manned rocket designed to land like a helicopter after carrying satellites into orbit was unveiled Monday. The Roton, built by Rotary Rocket Co. of Redwood City, is a launch vehicle powered by kerosene instead of costly hydrogen, which the firm hopes will cut launch costs by 90 percent. It has been designed to launch like a rocket, then deploy a propeller and land like a helicopter. ["New satellite launcher unveiled," Florida Today, March 2, 1999, p 2A.]

MARCH 2: Florida's space agency is asking the Legislature to significantly increase its funding or risk having the state lose its grasp on the commercial space market. Spaceport Florida Authority wants lawmakers to allocate $43.8 million in fiscal year 2000 to pay for various projects. The amount is far above the $450,000 to $750,000 in operating money the agency has received in each of the past few years. The authority is asking the Legislature to spend: * $15 million on a new Mission Management Center at Cape Canaveral Air Station or NASA's Kennedy Space Center. The center would house launch-control systems and accommodate companies that bring multimillion-dollar satellites to Florida for launch. An additional $10 million will be sought from the federal government and industry to complete the facility, which would replace a center built at Cape Canaveral during 1950s. * $15 million for an Academic Support Center that would include classrooms, offices and laboratories for university and college space-research and education programs. The facility would be at a new NASA-Spaceport Florida research park to be built on 400 acres of undeveloped land at Kennedy Space Center. * $8 million to pave roads and hook up water, sewer, electricity and communications equipment at the new research park. * $1.6 million to enlarge a new $8 million hangar, Tarmac and taxiway being built by the state to support next-generation launch vehicles. NASA is already kicking in $4.85 million to the hangar project. The expansion would enable the state to accommodate larger spaceships. *$900,000 for agency operation and administration expenses, money that would allow the authority to expand a staff of 15 to 16 or 17 full-time workers. *500,000 to develop a five-year spaceport capital improvement plan, and $200,000 to develop a plan to use Homestead Air Reserve Base south or Miami as a commercial site. ["Spaceport seeks boost from state," Florida Today, March 3, 1999, p 1A & 2A.]

◆ An undeveloped patch of land at Kennedy Space Center might someday become the Silicon Valley of space-related research if Spaceport Florida Authority's ambitious plan flies with lawmakers. The agency and KSC are planning a 400 acre research park for government, commercial and academic programs that would be located off State Road 3. The first project will be a $30 million space research and education complex. Work is expected to start late this year or early next year and be completed in 2001. It would have two parts: *A $15 million KSC lab to develop and process life sciences and microgravity experiments for the International Space Station and VentureStar, a next-generation space shuttle under development by Lockheed Martin. Spaceport plans to finance construction of the lab then lease it to KSC or its contractor. *$15 million state-funded academic facility that includes classrooms, offices and labs for university and community college research and education program. That project, which the Legislature has yet to back, would be designed to support NASA, military and industrial programs at the park. ["Land could develop into 400-acre research park," Florida Today, March 3, 1999, p 2A.]

MARCH 3: The launch of NASA's Wide-Field Infrared Explorer (WIRE) spacecraft aboard an Orbital Sciences Pegasus XL vehicle was postponed earlier this week. The new launch date is no earlier than March 4. Orbital Sciences engineers are troubleshooting the mechanism which secures the rudder pin on the Pegasus rocket and the device which retracts the pin for flight. The Pegasus is to be dropped from an L-1011 aircraft over the Pacific Ocean approximately 100 miles offshore from Vandenberg AFB, Calif. ["NASA spacecraft launch from VAFB postponed until March 4 at earliest," **KSC Countdown**, March 4, 1999.]

MARCH 4: The Wide-Field Infrared Explorer (WIRE) spacecraft was launched Thursday night over the Pacific Ocean. ["Satellite launch successful," <u>Florida Today</u>, March 5, 1999, p 2A.]

MARCH 5: A $54 million NASA spacecraft is tumbling through space today as ground controllers scramble to save the inaugural mission in a program aimed at determining whether life exists elsewhere. Although mission managers haven't declared the satellite dead yet, the craft was in critical condition Friday. Launched into space Thursday from off the coast of California by an air-launched Pegasus rocket, the Wide-Field Infrared Explorer spacecraft is equipped with a 560-pound telescope designed to study the formation of stars, planets and galaxies. Engineers think the WIRE mission began to go awry when a telescope cover inadvertently popped off after the craft reached an orbit 335 miles above Earth. Engineers think pent-up hydrogen gas then began shooting out spacecraft vents, sending the craft into a tumble. ["Star-studying satellite mission to be cut short," <u>Florida Today</u>, March 6, 1999, p 1A & 2A.]

MARCH 6: With NASA's $3 billion Hubble Space Telescope uncomfortably close to a shutdown in orbit, NASA officials this week are expected to approve an emergency repair flight in October. The objective is to replace a troubled system of gyroscopes that could fail any day, making it impossible to accurately point the 13-ton telescope at planets, stars, galaxies and other targets. The mission likely would fly on shuttle Discovery, and the work would be done by four spacewalkers who are busy preparing for the effort. NASA already had planned to repair the pointing system on a servicing flight scheduled for launch June 8, 2000. But serious consideration is being given to splitting the June 2000 mission in half. Under that scenario, a repair crew would be launched in October to fix the gyros and replace several other telescope parts. Mounting a mission in seven months would be daunting because NASA normally allots 12 to 15 months to prepare for a shuttle flight. But officials said the agency already has a leg up on much of the work that would be required for an October mission. ["NASA likely to OK Oct. shuttle flight to fix Hubble," <u>Florida Today</u>, March 7, 1999, p 1A & 2A.]

MARCH 8: In recent days, the National Aeronautics and Space Administration has been discussing a proposal to send the first crew of all women into orbit. The agency's hottest rumor was the subject of a Monday afternoon teleconference between officials at Johnson Space Center and NASA Headquarters in Washington. "I don't want to say there will never be a mission with all women, but crews aren't chosen that way," said Kathryn Clark, the chief scientist for the international space station program. Such a mission would be unprecedented. Only once has a manned spaceship departed Earth without a male passenger. That was in June 1963, when Soviet cosmonaut Valentina Tereshkova blasted off alone on a three-day flight. Air Force Lt. Col. Eileen Collins is set to become NASA's first female commander during a five-day shuttle Columbia mission scheduled for July. The resulting publicity from a flight of all women would likely rival that lavished on Mercury astronaut John Glenn's return to space last October. Concerns about the scientific merits will probably keep the mission from becoming a reality. Proponents of an all-women flight have suggested the mission could focus on gender-specific health issues and enlarge the existing body of knowledge on female astronauts. Studies might also compare the way women work together in space with how all-male and mixed crews operate. However, scientists have yet to identify any physiological changes from weightlessness that affect only one gender. ["An all-women shuttle mission? That idea might not fly," The Orlando Sentinel, March 9, 1999, p A-9.]

MARCH 9: Launched in 1977, both spacecraft Voyager 1 and Voyager 2 are healthy and continuing to explore at the edge of the solar system. Voyager 1 is 6.8 billion miles from Earth, traveling northward from Earth's vicinity at 38,718 mph; Voyager 2 is 5.3 billion miles from Earth, traveling southward about 35,500 mph. Voyager 1 is expected to reach the heliopause – the theoretical dividing line between our solar system and interstellar space – after 2001. ["Voyager spacecraft still going strong on way to edge of solar system," KSC Countdown, March 9, 1999.]

◆ A powerful rocket that promises to bring more commercial space business to Florida was wheeled to its launch pad Tuesday. The Lockheed Martin Atlas 3 is scheduled for launch June 15, carrying a U.S. television satellite. When it flies, it will do so using an engine designed and built in Russia – a first for the American space program. With the engine, the rocket will be able to carry communications satellites weighing as much as 9,000 pounds from Cape Canaveral Air Station. That's about 1,000 pounds more than its predecessors currently in service. Telstar 7 will be used to beam cable television, pay-per-view programming and other services to North America. ["Herculean Atlas 3 heralded," Florida Today, March 10, 1999, p 1B.]

MARCH 10: After weeks of intense planning, it's official. NASA announced Wednesday it will launch an emergency shuttle repair flight to the Hubble Space Telescope this fall to prevent the shutdown of the $3 billion observatory. NASA is targeting the launch of the emergency mission for mid-October, although the flight might be moved up or back a couple of weeks. The nine-day mission calls for three spacewalks to repair the gyroscopes and install a new main computer. The work will be done aboard shuttle Discovery. NASA already had planned to launch a servicing mission to Hubble in June 2000. The other chores tabbed for the June 2000 flight will be put off until sometime between December 2000 and March 2001. NASA experts think there is an 80 percent chance one of the working gyroscopes might fail before June 2000. ["Oct. Hubble-repair flight set," <u>Florida Today</u>, March 11, 1999, p 1A & 2A.]

MARCH 11: After years of delays on the International Space Station, the United States and Russia are working on a surprise plan to launch the first crew in October, three months earlier than planned, officials said Thursday. The unannounced change, if approved at a meeting next month, would represent a major publicity coup for the $50 billion station, which has been marred by repeated Russian delays in building the living quarters. The three-man Russian-American crew was scheduled to go up in January, but the latest idea is to launch two men a few weeks after the living-quarters module goes into orbit. The third man would go up by U.S. space shuttle soon after. Mike Baker, Moscow-based deputy head of the U.S. Johnson Space Center, said he thought an October launch would be approved, although it was now only under discussion. He said the early launch idea rose from contingency plans to send up astronauts if the living quarters do not properly attach to the rest of the station. ["U.S., Russia crew may launch early," <u>The Orlando Sentinel</u>, March 12, 1999, p A-3.]

◆ Space Shuttle Status Report, Thursday, March 11, 1999. STS-96: Yesterday, workers completed installation of the STARSHINE secondary payload into Discovery's cargo bay and interface verification tests are in work. Instrumentation has been installed on the orbiter to support International Space Station replenishment activities on orbit. This modification enables the Shuttle to provide ISS with oxygen, nitrogen and water. Today, technicians will install Discovery's drag chute. The controller for auxiliary power unit No. 2 is installed and undergoing checks this week. Installation of the integrated vehicle health monitoring system for the orbiter's main propulsion system continues on schedule and base heat shield fastener modifications continue. Booster stacking operations are complete in VAB high bay 3 and closeouts are ongoing. STS-93: Columbia is jacked and leveled in VAB high bay 2 undergoing system observation during a temporary storage period. Columbia will remain in the VAB until mid-April, when Shuttle Discovery rolls out of OPF bay 1. Columbia will then be transferred to OPF bay 1 to complete STS-93's orbiter pre-launch preparations. STS-99: Removal of Endeavour's three auxiliary power units is complete and installation of the left hand radiator No. 4 concluded this week. Valve modifications on freon coolant loop No. 1 continue to go well. Water spray boiler checks are under way. Preparations are in work to drain the orbiter maneuvering system crossfeed lines. STS-101: Work on Atlantis' tactical air and navigation system (TACAN) continues. Atlantis' orbiter maneuvering and reaction control systems are undergoing drain line modifications and the orbiter is scheduled to be powered up by the end of next week. Bruce Buckingham. (1999). **Kennedy Space Center Space Shuttle Status Report** [Online]. Available E-mail: domo@news.ksc.nasa.gov/subscribe shuttle-status [1999, March 11].]

MARCH 16: A NASA spacecraft is headed for the closest-ever encounter with an asteroid after successfully completing a crucial engine firing Tuesday. Its next stop: a July rendezvous with a space rock called 1992 KD to take pictures and study the remnants of the solar system's formation. The Deep Space 1 craft was launched from Cape Canaveral in October on a $152 million mission to test a dozen new technologies, including a futuristic ion engine that runs on charged atoms. The engine fired at 2:15 a.m. Tuesday, allowing the craft to begin to close the 90 million miles between it and the asteroid. NASA plans to use an ion engine to power a mission set for launch in 2003, when a probe is to explore a comet by landing on its surface. ["Spacecraft zips toward close encounter with asteroid," **Florida Today**, March 17, 1999, p 1A..]

◆ Chandra Status Report, Tuesday, March 16, 1999. In the Vertical Processing Facility (VPF) at KSC, work continues on schedule toward the launch on July 9 of the Chandra X-ray telescope aboard Space Shuttle Columbia on mission STS-93. The reworked Command and Telemetry Unit (CTU) arrived at KSC on March 2 and was reinstalled on Chandra the following day. The Interface Unit (IU) was received and reinstalled on March 8. A three-day retest of both elements was completed on March 11. Both these elements are responsible for handling commands and telemetry to and from the ground. This week, an overall Chandra state of health test is scheduled to be performed on March 17 and the attitude control thrusters will be helium flow-tested the following day. The solar arrays arrived March 9 at the Skid Strip on Cape Canaveral Air Station by C-17 aircraft. The three day effort to install and test the arrays is scheduled to begin on March 23. One of the two solar arrays will be deployed for a mechanical check and some limited work on March 27. On March 31, Chandra will be hoisted from its test stand onto a fueling stand. Loading of fuel into the spacecraft is currently planned to begin on April 5 and will take ten days to complete. These bi-propellant hydrazine and nitrogen tetroxide fuels will be used by the telescope to achieve its final orbit. Hydrazine will also be used by one of the telescope's subsystems associated with pointing of the telescope. Finally, the spacecraft's batteries will be installed on April 17. Chandra will then be ready for the arrival at the VPF of the Inertial Upper Stage Booster on April 19. Chandra arrived at KSC's Shuttle Landing Facility aboard an Air Force C-5 airplane on Feb. 4. Bruce Buckingham. (1999). **KSC Chandra X-Ray Telescope Status Report** [Online]. Available E-mail: domo@news.ksc.nasa.gov/subscribe shuttle-status [1999 March 16].]

MARCH 17: In a "what if" simulated rescue mission on Wednesday, March 17, the KSC response team trained for the unlikely scenario of a Shuttle mishap at the Shuttle Landing Facility. The Mode 7 simulation of an astronaut rescue exercised all aspects of command and control, search and rescue, and medical procedures required for a successful rescue. ["KSC rehearses for unhoped-for mishap at Shuttle Landing Facility," **KSC Countdown**, March 18, 1999.]

◆ The reworked Command and Telemetry Unit and an Interface Unit have been installed on the Chandra X-ray Telescope and successfully tested. Both elements are responsible for handling commands and telemetry to and from the ground. An overall state-of-health test began on March 17. The next phases will be testing the attitude control thrusters and installing and testing the solar arrays that arrived March 9 at Cape Canaveral Air Station. ["Payload update," **KSC Countdown**, March 18, 1999.]

MARCH 18: Despite objections from its own safety experts, the Air Force is moving with new momentum to radically change the way the public is protected from rocket accidents. The topic may come to a head Monday (March 22) during a video conference in Florida, Colorado and California to discuss the proposals, which are outlined in internal Air Force documents obtained by *Florida Today*. Sweeping in nature, the changes could strip independent authority from safety officers who blow up wayward rockets and cut hundreds of jobs at Patrick Air Force Base and Cape Canaveral Air Station. The discussions are ongoing, even though Air Force safety officials "strongly oppose" major parts of the plan, the documents say. Officials with the Federal Aviation Administration also do not want changes made until a review is completed late this year by the National Academy of Sciences. Air Force Col. Kevin Chilton, deputy director of operations at Air Force Space Command in Colorado Springs, Colo., said no changes will be made without careful consideration. Space Command is the organization pushing the changes. ["Rocket range changes on fast track," <u>**Florida Today**</u>, March 19, 1999, p 1A & 2A.]

◆ Space Shuttle Status Report, Thursday, March 18, 1999. STS-96: Yesterday, workers completed installation of Discovery's three main engines. Functional tests of the orbiter's global positioning system are also complete. Drag chute sensor instrumentation modification is in progress. Auxiliary power unit leak and functional checks are under way and installation of the integrated vehicle health monitoring system for the orbiter's main propulsion system continues. Engine heat shield fastener modifications continue as well. The rest of this week, workers will perform ammonia servicing and install the orbiter's transfer tunnel. STS-93: Columbia is jacked and leveled in VAB high bay 2 undergoing system observation during a temporary storage period. Columbia will remain in the VAB until mid-April, when Shuttle Discovery rolls out of OPF bay 1. Columbia will then be transferred to OPF bay 1 to complete STS-93's orbiter pre-launch preparations. STS-99: Installation of Endeavour's three auxiliary power units (APU) concludes today when technicians install APU No. 2. Valve modifications on freon coolant loop No. 1 continue to go well. Next week, fuel cells No. 1 and No. 2 are being replaced and preparations to replace the left-hand orbital maneuvering system engine are ongoing. STS-101: Work on Atlantis' tactical air and navigation system (TACAN) is nearing completion and today technicians begin work with a powered orbiter. Atlantis' orbital maneuvering and reaction control systems are undergoing drain line modifications. Bruce Buckingham. (1999). **Kennedy**

Space Center Space Shuttle Status Report [Online]. Available E-mail: domo@news.ksc.nasa.gov/subscribe shuttle-status [1999, March 18].]

MARCH 21: Florida soon will know where it stands in a national battle to lure a new, reusable spaceship that could dominate the next century's commercial launches. Within three months, Lockheed Martin will tell 15 states if proposals they submitted last year match plans for VentureStar - a spaceplane that could drive NASA's space shuttles and throwaway rockets out of business. The notice will come in the form of a briefing that spells out areas in which the states either meet or don't meet requirements for a spaceport to launch, land and service VentureStar. Lockheed Martin plans to invest $5 billion to build a two-ship VentureStar fleet and two spaceports to be located in the continental U.S. ["Florida faces VentureStar hurdle," **Florida Today**, March 21, 1999, p 1A & 2A.]

MARCH 23: With the Air Force considering significant changes to its Cape Canaveral launch operations, Congress today will start looking at ways to modernize archaic equipment used to monitor rocket flights. In a hearing before the House Science Committee, officials from the military, NASA and private industry are to explain why at least $3 billion must be spent to replace technology that dates to the 1960s. The topic is one part of a broad Air Force plan to make Cape Canaveral Air Station a cheaper and more efficient place for companies to launch commercial satellites into space. Other parts of the plan under discussion include changes in safety procedures used to launch spacecraft and destroy wayward rockets. The elimination of hundreds of jobs also is possible. ["Congress considers updating flight equipment," **Florida Today**, March 24, 1999, p 1B.]

MARCH 25: Space Shuttle Status Report, Thursday, March 25, 1999. STS-96: This week, the STS-96 crew completed orbiter and payload familiarization activities during a standard Crew Equipment Interface Test. Discovery's main propulsion system leak checks also concluded this week and now heat shield installation is under way. Instrumentation wiring continues in the orbiter's aft compartment as part of drag chute and vehicle health monitoring work. Auxiliary power unit functional testing is also in progress. STS-93: Columbia is jacked and leveled in Vehicle Assembly Building high bay 2 undergoing system observation during a temporary storage period. Columbia will remain in the Vehicle Assembly Building until mid-April, when Shuttle Discovery rolls out of Orbiter Processing Facility bay 1. Columbia will then be transferred to Orbiter Processing Facility bay 1 to complete STS-93's orbiter pre-launch preparations. STS-99: Installation of Endeavour's three auxiliary power units is complete. Replacement of window No. 8 is also complete. Preparations are in work to remove the left orbital maneuvering engine next week. Valve modifications on freon coolant loop No. 1 continue and water spray boiler checks are under way. Fuel cells No. 1 and No. 2 are scheduled for replacement this week. STS-101: Checks on Atlantis' main propulsion system liquid oxygen and liquid hydrogen valves are complete. Orbital maneuvering and reaction control systems are undergoing drain line modifications and helium system leak checks are ongoing. Bruce Buckingham. (1999).

Kennedy Space Center Space Shuttle Status Report [Online]. Available E-mail: domo@news.ksc.nasa.gov/subscribe shuttle-status [1999, March 25].]

MARCH 26: Chandra Status Report, Friday, March 26, 1999. In the Vertical Processing Facility (VPF) at KSC, work continues on schedule toward the launch on July 9 of the Chandra X-ray telescope aboard Space Shuttle Columbia on mission STS-93. Chandra successfully completed an overall state of health test on March 17 and the attitude control thrusters were helium flow-tested the following day without issues. The solar arrays were installed on March 26 and testing of the mechanical and electrical interfaces is underway. One of the solar arrays is to be fully deployed today or Saturday as part of a mechanical check and some limited work. In parallel, the Payload Operations Procedure (POP) test is underway which verifies the network command telemetry path from the telescope through the Deep Space Network to the ground control station at Harvard University. On March 31, Chandra will be hoisted from its test stand onto a fueling stand. Loading of fuel into the spacecraft is currently planned to begin on April 5 and will take ten days to complete. These bi-propellant hydrazine and nitrogen tetroxide fuels will be used by the telescope to achieve its final orbit. Hydrazine will also be used by one of the telescope's subsystems associated with pointing the telescope. Finally, the spacecraft's batteries will be installed on April 17. Chandra will then be ready for the arrival at the VPF of the Inertial Upper Stage Booster on April 19. Chandra arrived at KSC's Shuttle Landing Facility aboard an Air Force C-5 airplane on Feb. 4. Bruce Buckingham. (1999). **KSC Chandra X-Ray Telescope Status Report** [Online]. Available E-mail: domo@news.ksc.nasa.gov/subscribe shuttle-status [1999, March 26].]

MARCH 27: A Ukrainian rocket blasted off from a converted oil platform in the Pacific Ocean on Saturday night, launching new competition to Florida's commercial launch business. The Zenit rocket lifted off at 8:30 p.m. EST, and carried a demonstration satellite into orbit. About 14 minutes into the flight, controllers briefly lost contact with the satellite. However, communications were restored late Saturday night and the satellite was heading toward orbit, officials said. The international venture called Sea Launch is led by Boeing Corp. ["First sea-launched satellite lifts off in Pacific," <u>Florida Today</u>, March 28, 1999, p 1A.]

APRIL

APRIL 1: The White House has launched a multiagency review into what should be done about the growing obsolescence of U.S. space-launch ranges at Cape Canaveral and in California. The announcement late Wednesday said the National Security Council and the White House Office of Science and Technology Policy will lead the review. Along with NASA, the Department of Defense, Commerce and Transportation will participate. The effort reflects concern that space efforts are hampered by outdated equipment and procedures at the Cape and at Vandenberg Air Force Base in California. The review will examine how the government and private sector might share responsibility for upgrading the ranges and managing them in the future, White House officials said. The Kennedy Space Center, where space shuttles are processed and launched, is owned by National Aeronautics and Space Administration but uses the services of the Air Force's Cape Canaveral range. Both the Cape and Vandenberg are plagued by old computer, communications and rocket-tracking systems. Much of the equipment still be used dates to the early days of the space program, in the 1950s and 1960s. But the launch sites are under stress from a boom in private communications-satellite launches, as well as demand for more precision and operating flexibility from NASA and the military. The White House announcement did not suggest what might be done beyond Air Force plans to spend $1.2 billion in range upgrades by 2006. It said the review would begin this month and produce "interim recommendations" in time for formulation of the 2001 federal budget. That might not be fast enough for Congress, which is expected to hold more hearings and write legislation on the subject this year. ["Feds probe old space equipment," The Orlando Sentinel, April 2, 1999, p A-8.]

◆ NASA's Far Ultraviolet Spectroscopic Explorer (FUSE) satellite, which arrived April 1 at Hangar AE, Cape Canaveral Air Station, began processing on Monday, April 5. The protective shipping cover has been removed and the satellite is undergoing a functional test of its systems. Installation of the flight batteries and solar arrays will follow. Tests are also scheduled for the communications and data systems linking FUSE with the spacecraft control center at The Johns Hopkins University, Baltimore, Md. FUSE will investigate the origin and evolution of hydrogen and deuterium, the lightest elements in the universe. It will also examine the forces and process involved in the evolution of the galaxies, stars and planetary systems by investigating light in the far ultraviolet portion of the electromagnetic spectrum. ["FUSE satellite arrives at CCAS – processing begins this week," KSC Countdown, April 6, 1999.]

APRIL 3: EG&G Florida Inc. has cut off health care benefits to retired former employees. EG&G retirees say it's the first time this has had happened at the Space Center. NASA officials said they couldn't recall it happening before. EG&G cut off the benefits after the company lost its contract with NASA to provide launch support services at the Space Center. The company held the contract for 15 years. EG&G

management says its collective bargaining agreements with employees made it clear the Wellesley, Mass.-based company would end benefits if it lost the NASA contract. ["Retirees fight for benefits as companies cut costs," <u>Florida Today</u>, April 4, 1999, p 1E & 4E.]

APRIL 5: Space Shuttle Status Report, Monday, April 5, 1999. STS-96: Over the weekend, workers completed leak checks on Discovery's hydraulic power system and crew cabin. Water spray boiler servicing is under way and checks of the integrated vehicle health monitoring system continue. Orbiter aft and midbody compartment close-outs are in progress. External tank and solid rocket booster close-outs are ongoing in the Vehicle Assembly Building. Later this week, workers will vent the external fuel tank's thermal foam by drilling small holes in the foam. This effort will help reduce the amount of orbiter-damaging foam debris generated during ascent. STS-93: Columbia is jacked and leveled in Vehicle Assembly Building high bay 2 undergoing system observation during a temporary storage period. Columbia will remain in the VAB until mid-April, when Shuttle Discovery rolls out of Orbiter Processing Facility bay 1. Columbia will then be transferred to OPF bay 1 to complete STS-93's orbiter prelaunch preparations. STS-99: Preparations to replace Endeavour's left orbital maneuvering system engine are complete and removal efforts begin today. Fuel cells No. 1 and No. 2 are being replaced this week as well. Valve modifications on freon coolant loop No. 1 continue and water spray boiler checks are under way. STS-101: Atlantis' main propulsion system is undergoing leak and functional tests this week. The oxygen tank for the orbiter's power reactant storage and distribution system has been removed to accommodate the replacement of an attach bolt. The tank will be reinstalled tomorrow. Later this week, workers will remove the external airlock to gain additional orbiter access. Bruce Buckingham. (1999). **Kennedy Space Center Space Shuttle Status Report** [Online]. Available E-mail: domo@news.ksc.nasa.gov/subscribe shuttle-status [1999, April 5].]

APRIL 6: Chandra Status Report, Tuesday, April 6, 1999. In the Vertical Processing Facility (VPF) at KSC, work continues on schedule toward the launch on July 9 of the Chandra X-ray telescope aboard Space Shuttle Columbia on mission STS-93. The solar arrays were installed on March 26 and testing of the mechanical and electrical interfaces was successfully finished on April 1. Their high performance could be noticed when during the deployment test of one of the arrays a small amount of power was already being generated with only the available lighting of the VPF. On April 2, Chandra was hoisted from its test stand onto a fueling stand. Preparations for fueling are now underway. Loading of the nitrogen tetroxide oxidizer into the spacecraft is currently planned to occur on April 9. Then the bi-propellant hydrazine, used by the telescope's propulsion system to achieve final orbit, will be loaded aboard April 13. Hydrazine, used by a subsystem that contributes to pointing the telescope, will be loaded April 15. Finally, the spacecraft's batteries will be installed. Chandra will then be ready for the arrival at the VPF of the Inertial Upper Stage Booster on April 19 as scheduled. Chandra arrived at KSC's Shuttle Landing Facility aboard an Air Force C-5 airplane on Feb. 4. Bruce Buckingham. (1999). **KSC Chandra X-Ray Telescope Status**

Report [Online]. Available E-mail: domo@news.ksc.nasa.gov/subscribe shuttle-status [1999, April 6].]

◆ Boeing's Delta 3 rocket's planned launch Tuesday night was scrubbed for the second straight day, forcing officials to move its next attempt to no earlier than mid-April. Tuesday's attempt was called off after several delays caused by different technical issues. The attempt was ultimately canceled because of a problem with Air Force equipment used to track the rocket in flight. An attempt Monday was canceled because winds could have allowed a toxic cloud to drift toward populated areas if the rocket exploded. ["Equipment problems delay launch of Boeing's Delta 3," <u>Florida Today</u>, April 7, 1999, p 1A.]

◆ There are three new attractions at the Kennedy Space Center Visitor Complex. "We're taking the KSC Visitor Complex to the next level," said Rick Abramson, president and chief operating officer of Delaware North Park Services, the company that runs the complex for NASA. The complex's three new attractions are part of an ongoing renovation plan that totals $100 million to date. They include: *Robot Scouts – a walk through exhibit that features "Starquester," a talking robot who guides visitors through some of NASA's many explorations of the solar system. *"Quest for Life" – a 15 minute movie that focuses on the search for microbes on Mars, planets in other solar systems and the potential for an ocean on Jupiter's moon Europa. *Merritt Island National Wildlife Refuge – a wooden walk-through exhibit displaying the foliage and animals that share the land with Kennedy Space Center. ["KSC launches attractions," <u>Florida Today</u>, April 9, 1999, p 1B.]

APRIL 8: Space Shuttle Status Report, Thursday, April 8, 1999. STS-96: Shuttle managers are now targeting May 20 for the launch of Shuttle Discovery on mission STS-96. That date on the Eastern Range became available after Delta FUSE managers announced a slip in their launch date. Shuttle processing managers at KSC have been protecting a May 20 launch option and are ready to support. Discovery's Ku band system interface testing is complete. Auxiliary power unit lubrication servicing is under way today and checks of the integrated vehicle health monitoring system continue. Orbiter aft and midbody compartment close-outs proceed on schedule. External tank and solid rocket booster close-outs continue in the Vehicle Assembly Building and workers have completed thermal foam venting on the external tank. STS-93: Columbia is jacked and leveled in Vehicle Assembly Building high bay 2 undergoing system observation during a temporary storage period. Columbia will remain in the VAB until April 14, when Shuttle Discovery rolls out of Orbiter Processing Facility bay 1. Columbia will then be transferred to OPF bay 1 to complete STS-93's orbiter prelaunch preparations. STS-99: Removal of Endeavour's left orbital maneuvering system (OMS) engine is complete. Fuel cells No. 1 and No. 2 have been replaced. Leak checks on freon coolant loop No. 1 revealed an alternate power converter unit leak and technicians are currently troubleshooting the issue. Preparations are now in work for next week's removal of the right hand OMS pod.

STS-101: Atlantis' external airlock has been removed to provide workers additional access to the orbiter's midbody. Main propulsion system leak and functional tests continue. Replacement of the strut and attach bolt for the power reactant storage and distribution system oxygen tanks is ongoing. Bruce Buckingham. (1999). **Kennedy Space Center Space Shuttle Status Report** [Online]. Available E-mail: domo@news.ksc.nasa.gov/subscribe shuttle-status [1999, April 8].]

APRIL 9: A $250 million military satellite was ferried into space Friday aboard an Air Force Titan 4 rocket launched from Cape Canaveral Air Station. The launch went flawlessly, restoring the nation's only means of getting large pentagon spy satellites into orbit. The satellite will use its thrusters to move into its final position about 22,300 miles above Earth. From there it will a join a fleet of spacecraft called Defense Support Program satellites that use infrared sensors to instantly detect missile launches and nuclear detonations worldwide. ["Titan 4 'back on track'," **Florida Today**, April 10, 1999, p 1A.]

APRIL 10: The nation's newest missile-warning satellite is circling in a useless orbit today and might be a total loss, Air Force officials said Saturday. Launched Friday aboard an Air Force Titan 4 rocket, the satellite was hurled into the wrong orbit. The Titan 4 rocket performed flawlessly, but a Boeing-built upper stage, intended to boost the Defense Support Program spacecraft to it work station 22,300 miles above the Earth, failed to do so. It's unclear whether the satellite has enough fuel to fly into a useful orbit. ["Military satellite in useless orbit," **Florida Today**, April 11, 1999, p 1A.]

◆ Within the next two years, NASA may be flying all-female space shuttle crews – for science. With a new space station on the horizon and increasing talk of trips to Mars, NASA wants to make sure it protects the health of all its astronauts, male and female. NASA is seeking multiple second opinions to determine whether more gender-specific research is needed. It wasn't until the last few years that NASA could even consider putting together an all-female crew. Every shuttle mission requires two pilots, and NASA only recently added its second and third female shuttle pilots. The 119 current astronauts include 29 women. "Sooner or later it's going to happen" whether it's deliberate or not, said shuttle program director William Readdy. ["NASA ponders female crews," **Florida Today**, April 11, 1999, p 1A & 2A.]

APRIL 12: Space Shuttle Status Report, Monday, April 12, 1999. STS-96: Over the weekend, workers removed "D" hatch from Discovery's transfer tunnel adapter to accommodate structural and latch inspections. Workers replaced it with orbiter Endeavour's "D" hatch. Due to the additional work, managers now plan to transfer Discovery to the VAB on Thursday, but do not expect Discovery's arrival at the launch pad to be delayed. Orbiter aft and midbody compartment close-outs continue on schedule. Today, workers will close Discovery's payload bay doors and tomorrow orbiter weight and center of gravity tests are planned. STS-93: Columbia remains in

Vehicle Assembly Building high bay 2. It underwent final powered checkup in the VAB on Friday. Columbia will be transferred to OPF bay 1 to complete orbiter processing for mission STS-93 following Shuttle Discovery's move to the VAB. STS-99: Checks on Endeavour's left hand orbital maneuvering system (OMS) engine are in work. Fuel cells No. 1 and No. 2 are also undergoing checks following their recent replacement. Servicing of freon coolant loop No. 1 is under way and workers will remove Endeavour's right hand OMS pod tonight. STS-101: Replacement of the strut attach bolt for Atlantis' power reactant storage and distribution system oxygen tanks is complete. Main propulsion system gaseous hydrogen leak checks are ongoing. Bruce Buckingham. (1999). **Kennedy Space Center Space Shuttle Status Report** [Online]. Available E-mail: domo@news.ksc.nasa.gov/subscribe shuttle-status [1999, April 12].]

◆   A new communications satellite is in orbit today, ready to expand television and other services to the people of Turkey, Poland and the nations of North Africa. The 7,000-pound satellite was carried into orbit Monday by a Lockheed Martin Atlas rocket launched from Cape Canaveral Air Station. The $200 million mission is to deliver the satellite for the Paris-based Eutelsat company, which plans to use it for television, Internet and other telecommunications services. The Atlas took off at 6:50 p.m. from Cape Canaveral. ["Atlas lifts off without a hitch," <u>Florida Today</u>, April 13, 1999, p 1B.]

◆   NASA's planned launch this summer of a $1.5 billion telescope may be delayed because it uses the same rocket engine that apparently failed on a Pentagon spy satellite Friday. Set for launch July 9 on shuttle Columbia, the Chandra X-Ray Observatory will help scientists piece together information about how the universe works. The observatory will use an Inertial Upper Stage rocket engine made by Boeing Co. to boost it into orbit 86,000 miles above the Earth. The same kind of engine apparently failed Friday during an Air Force Titan 4 rocket mission. NASA officials said Monday they won't launch Chandra until they know what caused the accident. An investigation is under way and could take several months. ["Rocket engine failure may delay telescope launch," <u>Florida Today</u>, April 13, 1999, p 1A.]

APRIL 13: An attempt is expected to be mounted this weekend to salvage a piece of American space history – the sunken Mercury capsule of astronaut Gus Grissom. Two previous attempts to locate Liberty Bell 7 have failed, but a new expedition is planning to sail Saturday or Sunday from Port Canaveral, said Howard Benedict, executive director of the Astronaut Scholarship Foundation in Titusville. The Discovery Channel is to film the expedition for a TV special. The expedition is led by Curt Newport of Virginia. Newport is an underwater salvage expert who tried without success in 1992 and 1993 to locate the craft about 300 miles east of Cape Canaveral. Grissom became the second American in space on a 15-minute, suborbital flight July 21, 1961. After splashing down in the Atlantic Ocean, the capsule's hatch door blew off prematurely and water poured inside. Grissom almost drowned, and the capsule sank in about 15,000 feet of water. To this day, no one knows why the craft's door opened early. Its recovery may help solve the mystery. ["Salvagers hope to recover sunken space capsule," <u>Florida Today</u>, April 14, 1999, p 1A.]

APRIL 15: Space Shuttle Status Report, Thursday, April 15, 1999. STS-96: Shuttle orbiter Discovery rolled from Orbiter Processing Facility bay 1 to the Vehicle Assembly Building today at about noon. Tonight, the vehicle will be lifted into VAB high bay 3 and then soft mated to the STS-96 external tank and solid rocket booster stack. Following final electrical and mechanical connections, Shuttle interface testing is scheduled for April 19. Discovery is targeted for rollout to the launch pad on April 21. At the pad, technicians will remove the two integrated electronic assembly (IEA) boxes from both solid rocket boosters. One IEA is in the forward assembly and one is in the aft assembly. The IEAs receive and execute SRB separation and range safety commands. Inside each IEA is a multiplexer demultiplexer that houses a printed circuit card. Recent tests have revealed that these cards may develop electrical shorts due to normal flight operations. As a precaution, Shuttle managers have decided to replace the suspect components in parallel with standard prelaunch work at the pad. No impact to the May 20 target launch date is expected. STS-93: Shortly after noon today, Columbia arrived at OPF bay 1 having completed its temporary storage period in VAB high bay 2. Orbiter processing for mission STS-93 will resume following routine receiving inspections. STS-99: Checks on Endeavour's orbital maneuvering system continue. Servicing of freon coolant loop No. 1 is complete and preparations to service loop No. 2 are in work. This week, workers will install the orbiter docking system harness. STS-101: Following replacement of the strut attach bolt for Atlantis' power reactant storage and distribution system, leak checks of oxygen tanks are now complete. Payload bay liner retainer modifications are under way and main propulsion system testing continues this week. Bruce Buckingham. (1999). **Kennedy Space Center Space Shuttle Status Report** [Online]. Available E-mail: domo@news.ksc.nasa.gov/subscribe shuttle-status [1999, April 15].]

◆ Moving into the new millennium, Atlantis is the first orbiter to be upgraded with a Multifunction Electronic Display Subsystem (MEDS), known as the glass cockpit. The full-color flat panel MEDS replaces electromechanical cockpit displays like cathode ray tube screens, gauges and instruments. The new system improves crew/orbit interaction with easy-to-read graphic portrayals of key flight indicators like attitude display and Mach speed. ["Shuttle Atlantis shows off new cockpit display that provides easy use, crew interaction," **KSC Countdown**, April 15, 1999.]

APRIL 16: An Air Force investigation into a botched satellite launch could mean another delay for NASA's $1.5 billion Chandra X-ray telescope. The Air Force has impounded an upper-stage motor intended for the telescope, to be launched in July aboard space shuttle Columbia. It's the same kind of motor that apparently malfunctioned April 9 and left the military's newest Defense Support Program satellite stranded in a useless orbit. The National Aeronautics and Space Administration is working with the Air Force's accident investigation board, which convened earlier this week. The Air Force must deliver the impounded Boeing-built motor to NASA by the end of April if the space agency hopes to meet its July 9 launch date for Chandra, payload manager Scott Higginbotham said. The telescope-delivery mission already is a year late because of bad circuit boards that had to be replaced in Chandra and other problems. Chandra, a Hubble-caliber observatory, should have rocketed into orbit last summer. Unlike Hubble, the X-ray telescope will be placed in an orbit too high to be reached by spacewalking astronauts for repair. ["NASA launch faces new threat," The Orlando Sentinel, April 17, 1999, p A-21.]

◆ The family of astronaut Gus Grissom won't be rooting for an expedition that plans to sail Sunday in search of his sunken capsule, the Liberty Bell 7. "I hope they do not find it," Betty Grissom, the astronaut's widow, said Friday from her home in Houston. Financed by the Discovery Channel, the expedition will begin Sunday on a 10-day trip to locate and retrieve the capsule that has been under 3 miles of water since 1961. ["Grissom's widow no fan of plan to raise capsule," Florida Today, April 17, 1999, p 1A.]

APRIL 18: A bill pending in the Florida Legislature would encourage companies interested in building new space vehicles and other projects to do so not just at Cape Canaveral and Kennedy Space Center but at Eglin Air Force Base. The rationale is that the sprawling Panhandle installation would give the state another card to play in efforts to lure more space business to Florida. "Anything we can do to make space a high priority in Florida, we need to do," said bill sponsor Rep. Howard Futch, R-Melbourne Beach. Spaceport Florida Authority, the quasi-government agency formed a decade ago to help bring aerospace business to Florida, has been working hard to do that at Cape Canaveral. It has built its own launch pad there and is constructing another that will allow companies to quickly launch payloads into orbit. Now the agency is pushing a bill sponsored by Futch and similar legislation by Sen. Jim Sebesta, R-St. Petersburg, that would extend its territory to Eglin. The idea is to work out a deal with the Air Force that would allow potential manufacturers of next-generation reusable spacecraft to launch and land the vehicles from one of the base's many runways. Facilities at Cape Canaveral Air Station and Kennedy Space Center also are being touted for projects, such as Lockheed Martin's VentureStar spaceship, which someday could replace NASA's space shuttles. ["State may use Eglin to lure space business," Florida Today, April 19, 1999, p 1A & 5A.]

APRIL 19: An underwater salvage team set off Monday in search of Mercury astronaut Gus Grissom's space capsule, entombed in the Atlantic for 38 years. A 180-foot ship, the Needham Tide, headed for three-mile deep waters 300 miles southeast of Cape Canaveral. The ship was a day late raising anchor because the crew needed to replace a switch on a hydraulic power unit. ["Salvage team sets out to find space capsule," Florida Today, April 20, 1999, p 2B.]

◆ It has been called boring, worthless, silly and foolish – a pointless, $75 million screen saver. But despite groans and complaints from Congress and others, NASA is forging ahead with Vice President Al Gore's pet project: a space camera that will broadcast via the Internet a constant view of the Earth sailing along its primordial orbit 93 million miles from the sun. Named Triana, after the Spanish sailor credited with sighting the New World on Christopher Columbus' first voyage, Gore's satellite is to be launched from shuttle Columbia in December 2000. ["Gore craft set for launch in 2000," Florida Today, April 20, 1999, p 1A & 9A.]

APRIL 22: Boeing will have to wait until at least Saturday to try to launch its new Delta 3 rocket again, after stopping a launch attempt Thursday at the last moment. After a trouble-plagued countdown, the company ticked down to the final second when launch computers failed to send the command to ignite the engines. Officials did not know immediately what caused the last-minute abort, but expressed confidence they will try again soon. Thursday's launch attempt was the fourth this month. Three earlier tries, including one Wednesday, were canceled for a variety of technical or weather problems. The Delta 3 rocket has not flown since its maiden voyage ended in disaster last August, when a vehicle exploded and destroyed a U.S. television satellite. ["Glitch postpones Delta 3 launch," Florida Today, April 23, 1999, p 1A.]

APRIL 23: NASA is getting ready to launch its first space shuttle in six months. The agency moved shuttle Discovery to its Kennedy Space Center launch pad Friday in preparation for its planned May 20 liftoff to the International Space Station. The last time a shuttle flew was early December when six astronauts aboard Endeavour began building the outpost. The current situation marks the longest the shuttle fleet has been idle since 1990, when dangerous hydrogen fuel leaks grounded the fleet. This time the blame falls on the Russians, who are far behind in getting a key station segment called the Service Module ready for launch. NASA thinks Russia may launch the module in September, although it could happen as late as November. As a result, Discovery's 10-day flight could be the only one to the station this year, way below the four station missions NASA had on the books in January. Despite the flight stoppage, NASA officials say the KSC work force has stayed sharp through repeated practice countdowns where all kinds of problems have been thrown their way. They'll have a chance to do another one next week when Discovery's crew come to KSC for a final prelaunch rehearsal. ["Shuttle back on launch pad," Florida Today, April 24, 1999, p 1A.]

◆ The launch of Boeing's new Delta 3 rocket has been postponed indefinitely while engineers investigate a last-second scrub Thursday night. Company officials still do not know why the rocket's computers failed to send a command to start the vehicle's engines as the countdown clock rolled to zero. At best, a new launch date may be announced early next week. ["Investigation postpones Delta launch," Florida Today, April 24, 1999, p 8A.]

APRIL 24: Calling it critical to the state's economy, Gov. Jeb Bush pledged Saturday to make the aerospace industry a top priority. On his first visit to Brevard County since taking office, Bush said the Space Coast has an "enthusiastic advocate" in the governor's mansion. He told aerospace leaders gathered at the annual Debus Award ceremony that he will guide the state to remain a force in the industry. Bush's remarks were made during the ceremony to honor this year's Debus Award winner, Ed O'Connor, director of Spaceport Florida Authority and 30-year veteran in the business. The award – named for Kennedy Space Center's first director, Kurt Debus – honors state leaders who have fostered the aerospace industry. It is given by the National Space Club. ["Bush promises to be aerospace advocate," **Florida Today**, April 25, 1999, p 1A.]

APRIL 25: NASA's space shuttle launches will be sold-out shows this year. The agency has cut back on the number of free car passes available to the public, and the remaining ones have been snapped up by sightseers who want to watch a launch from NASA Causeway on Merritt Island. "They're spoken for already for the entire year," said Kennedy Space Center spokeswoman Lisa Malone. The agency used to issue about 5,000 car passes for every shuttle flight, and anyone could get a pass by writing or calling KSC. Some of the passes also are distributed to NASA employees and contractors. But safety concerns have prompted the agency to cut the number in half. Officials say new studies show no more than 2,500 cars could be evacuated quickly enough from the causeway if a shuttle accident sent a toxic cloud over the area. As a result, fewer passes will be available for every launch after the scheduled flight of shuttle Discovery on May 20. ["Free launch passes all handed out," **Florida Today**, April 26, 1999, p 1A.]

APRIL 27: Space Shuttle Status Report, Tuesday, April 27, 1999. STS-96: Last Friday, Shuttle Discovery rolled out to Launch Pad 39B. Over the weekend, workers conducted routine launch pad validations and Shuttle main engine frequency response tests. The SPACEHAB payload is at the pad and will be installed in the payload changeout room later today. The payload will be installed in the orbiter tomorrow morning. The STS-96 flight crew arrived at KSC yesterday in preparation for this week's Terminal Countdown Demonstration Test. The crew will conduct orbiter and payload familiarization activities over the next few days and then participate in a launch dress rehearsal on Thursday. STS-93: NASA decided yesterday to postpone mating Chandra with its Inertial Upper Stage (IUS) pending the U.S. Air Force investigation into problems with the April 9 launch of a Department of Defense Satellite that utilized an IUS. This decision will likely delay launch of mission STS-93 beyond its current target launch date of July 9. NASA managers are participating in the Air Force investigation and have determined not to launch Mission STS-93 until the IUS situation is fully understood. Chandra is currently in KSC's Vertical Processing Facility. STS-99: Wiring modifications on Endeavour's radiator isolation valve continued over the weekend. Yesterday, technicians began reservicing the orbiter's repaired freon coolant loop No. 1. Orbiter docking system harness installation continues and external airlock installation concludes May 6. STS-101:

Atlantis is undergoing a wireless video modification. Main propulsion system leak and functional tests are in work and standard life support system leak checks continue. Payload bay liner modifications proceed on schedule. Bruce Buckingham. (1999). **Kennedy Space Center Space Shuttle Status Report** [Online]. Available E-mail: domo@news.ksc.nasa.gov/subscribe shuttle-status [1999, April 27].]

◆　The 36th Space Congress begins on Tuesday, April 27, at the Radisson Resort at the Port. Topics being addressed are the push to Mars and beyond, the military's changing role in space and commercial access to space. ["36th Space Congress focuses on military, commercial role in space," **KSC Countdown**, April 27, 1999.]

APRIL 30: An $800 million military communications satellite was launched into the wrong orbit Friday in the third straight failure for Air Force Titan 4 rockets. The accident, following losses two weeks ago and last August, raises serious questions about the nation's ability to get critical satellites into orbit. Standing 19 stories high, the Titan 4 blasted off from Cape Canaveral Air Station at 12:30 p.m. However, the rocket's upper stage apparently misfired. That stranded its payload – a Milstar communications satellite – in an egg-shaped orbit and high and low points of 3,100 and 4600 miles, respectively. Ground controllers were trying late Friday to salvage the spacecraft, but it probably doesn't have enough fuel to maneuver into its proper orbit. Brig. Gen. Craig Cooning, the Air Force's Milstar program director, said it was too soon to tell whether the satellite could be saved. The string of failures is expected to indefinitely ground the Titan rocket fleet. The planned May 7 launch of a Titan rocket from Vandenberg likely will be postponed while the Air Force investigates Friday's mishap. ["Titan fails again," <u>Florida Today</u>, May 1, 1999, p 1A & 2A.]

MAY

MAY 1: Kennedy Space Center recently started offering a wildlife tour of some of the 140,000 acres that surround the Vehicle Assembly Building and the launch pads. KSC uses only about 5 percent of the land owned by the federal government here. The rest of the rugged terrain – home to 310 bird species and 21 endangered animals – has been left untouched for years. The nature tours at SKC are offered twice daily on Wednesdays and Fridays and leave from the Visitors Center. ["Ecotour shows pristine KSC lands," <u>Florida Today</u>, May 2, 1999, p 1B & 2B.]

MAY 3: Lost at sea for 38 years, astronaut Gus Grissom's Mercury capsule was found during the weekend by an underwater salvage team that had been searching for the spacecraft 300 miles offshore. Liberty Bell 7 is 3 miles deep in the Atlantic. The cable to a remotely operated rover used to identify and photograph the capsule snapped in rough seas Saturday night and the rover sank. The salvagers are returning to Port Canaveral for new equipment and will have to wait at least a few weeks before going back to recover the spacecraft. The only U.S. spacecraft ever lost after a successful mission, the capsule is shiny in spots with an intact window and the name "Liberty Bell 7" clearly printed beneath it. Even the fake crack that was painted on the capsule to replicate the real Liberty Bell is visible, as are the singe marks left by the explosives that blew out the hatch following splashdown July 21, 1961. ["Grissom's lost space capsule found," <u>Florida Today</u>, May 3, 1999, p 1A.]

MAY 4: Boeing's Delta III rocket appeared to have suffered its second mishap in as many launches on Tuesday, launching a $145 million Asian television satellite into a lower than planned orbit. Flight controllers weren't sure what caused the problem, but data from the rocket suggested its second stage shut down early. It was unclear how high an orbit the Orion-3 satellite had reached. The one and only previous Delta III mission ended 70 seconds after liftoff on Aug. 26, when a guidance problem caused the rocket to explode. Liftoff of the long-awaited second mission finally came at 9 p.m. on Tuesday. It appeared to be a smooth ride to space. But about 30 minutes after launch, data from the second stage was lost shortly after the stage was set to reignite. ["Delta III fouls up another satellite," <u>The Orlando Sentinel</u>, May 5, 1999, p A-3.]

MAY 6: Space Shuttle Status Report, Thursday, May 6, 1999. STS-96: The SPACEHAB tunnel has been mated to Discovery inside the payload bay and leak checks are complete. The orbiter's prelaunch propellant loading began yesterday and will continue through Saturday. Tomorrow, auxiliary power unit No. 2 will be hot-fired as part of planned launch preparations. This coming weekend, workers will complete SPACEHAB payload testing and drag chute door installation. During yesterday's Flight Readiness review, Shuttle managers confirmed plans to use a functional drag-chute and strengthened drag chute door assembly on the upcoming flight. Inconel hinge-pins will replace the aluminum pins that failed during the STS-95 flight, allowing the drag chute door to fall off during liftoff. STS-93: Payload bay radiator inspections concluded yesterday. Final inspection of Columbia's thermal protection system is under way and auxiliary power unit lubrication oil servicing resumes tomorrow. Tomorrow, Columbia's drag chute door will be installed. The orbiter's nose and main landing gear is slated for installation next week. STS-99: Modifications to Endeavour's main engine dome heat shields continue. Freon coolant loop No. 2 modifications are in work also. Next week, fuel cell voltage testing begins. Orbiter docking system harness installation continues. STS-101: Atlantis continues to undergo wireless video modifications. Engine dome heat shield removal begins tonight. Navigation system wiring efforts are in work today and tomorrow. Payload bay liner modifications proceed on schedule. Bruce Buckingham. (1999). **Kennedy Space Center Space Shuttle Status Report** [Online]. Available E-mail: domo@news.ksc.nasa.gov/subscribe shuttle-status [1999, May 6].]

MAY 10: Space Shuttle Status Report, Monday, May 10, 1999. STS-96: Last Friday, workers completed a hot-fire test on Discovery's auxiliary power unit No. 2 and today APU pressurization is under way. Orbiter crew compartment purge testing concluded over the weekend. At the pad, the Rotating Service Structure was placed around the Shuttle on Friday and the payload bay doors are open. SPACEHAB interface verification testing concludes today and the payload bay doors will be closed for flight tomorrow afternoon. Equipped with new instrumentation, the drag chute door has been installed and will undergo tests this week. Mating of the orbiter midbody umbilical unit and subsequent leak checks are in work. Technicians are also inspecting the orbiter and external tank for possible hail damage from a weekend storm at KSC. STS-93: Inspections of Columbia's newly installed payload bay floodlights are complete. Functional checks of the orbiter maneuvering system and potable water servicing are in progress. Leak checks are in work on the exhaust ducts for auxiliary power units No. 1 and No. 3. The orbiter's nose and main landing gear is slated for installation this week. STS-99: Integration of Endeavour's hydraulic system and Shuttle main engines is complete. Fabrication and installation of tubing for freon coolant loop No. 2 continues. Orbiter docking system harness installation continues. Fuel cell voltage testing begins this week and wireless video modifications continue. STS-101: Atlantis continues to undergo wireless video modifications. Payload bay liner modifications proceed on schedule. Mass spectrometer leak checks on the orbiter's fuel cell power plant are in progress. Bruce Buckingham. (1999). **Kennedy Space Center Space Shuttle Status Report** [Online]. Available E-mail: domo@news.ksc.nasa.gov/subscribe shuttle-status [1999, May 10].]

MAY 10: Rain damaged a $40 million military satellite on its launch pad during last weekend's storms, adding another mishap to the Air Force's rash of troubles at Cape Canaveral Air Station. Attached to the top of a Delta 2 rocket, the satellite apparently got wet when water leaked into the facility that protects it during launch preparations, Air Force officials said Monday. They could not say how the water seeped in or how extensive the damage might be. As a result, the satellite's planned launch May 23 is postponed indefinitely when the craft is removed from the rocket and inspected. Brig.

Gen. Randall Starbuck said the Air Force will proceed methodically with repairs and an investigation into the mishap, which is the latest in a string of accidents at the Cape during the past nine months. The accident occurred Saturday (May 8), when workers were called off the launch pad structure because of lightning danger as thunderstorms rolled through Brevard County. The damaged satellite is the third in a series of 21 advanced GPS spacecraft planned to replace aging satellites already in orbit. ["Weekend's rain damaged satellite atop rocket at Cape," Florida Today, May 11, 1999, p 1A & 2A.]

MAY 11: Space Shuttle Status Report, Tuesday, May 11, 1999. STS-96: Orbiter midbody umbilical unit mating and leak checks are complete. Workers completed testing Discovery's Space to Space Orbiter Radio early this morning and the auxiliary power unit No. 2 tank has been pressurized. Drag chute door close-outs concluded yesterday and orbiter aft compartment close-outs are in work. Interface verification testing for the SPACEHAB payload is complete and the payload bay doors will be closed for flight this afternoon. An exposed fire alarm cable conductor was repaired last night on the Mobile Launcher Platform (MLP). Inspections of similar cables on the MLP are in work today. Following yesterday's orbiter and external tank inspections, workers confirmed that Discovery has not sustained any hail damage from a severe weekend storm. However, impact points were identified on the upper portion of the external tank's outer foam. Only 8 to 10 impact points require repair efforts, the largest measuring 2 = inches deep and about 2 inches wide. This week workers will gain access to the affected areas using a platform from the gaseous oxygen vent arm and complete the minor repairs with no significant impact to the work schedule. STS-93: Functional checks of Columbia's orbiter maneuvering system and potable water servicing continue. Leak checks continue on the exhaust ducts for auxiliary power units No. 1 and No. 3. Payload bay close-outs are scheduled to begin this week. STS-99: Fabrication and installation of tubing for Endeavour's freon coolant loop No. 2 continues. Orbiter docking system wire harness installation concludes tomorrow. Fuel cell voltage testing begins this week and wireless video modifications continue. STS-101: Atlantis continues to undergo wireless video modifications. Payload bay liner modifications proceed on schedule. Mass spectrometer leak checks on the orbiter's fuel cell power plant continue. TACAN antenna installation is in progress. Bruce Buckingham. (1999). **Kennedy Space Center Space Shuttle Status Report** [Online]. Available E-mail: domo@news.ksc.nasa.gov/subscribe shuttle-status [1999, May 11].]

◆ A House subcommittee will conduct hearings this month to investigate whether three consecutive Air Force rocket failures have injured U.S. military capabilities. Launched from Cape Canaveral Air Station, the three Titan 4 rocket accidents have cost taxpayers $3 billion and rendered useless a trio of advanced spy satellites. Calling the losses "unacceptable," Rep. Mike Castle, R-Del., said Tuesday he will convene the hearings before the House Technical and Tactical Intelligence subcommittee. He said the hearings likely will not be open to the public. ["House panel to probe rocket failures," Florida Today, May 12, 1999, p 1A.]

MAY 12: Space Shuttle Status Report, Wednesday, May 12, 1999. STS-96: SPACEHAB stowage concluded yesterday and loading of Discovery's Mass Memory Units is complete. Additional testing of the experimental Space to Space Orbiter Radio system resumes today requiring the orbiter's payload bay doors to remain open until tomorrow afternoon. Discovery's aft compartment doors will be installed Friday. Hail damage repair efforts are in progress on the upper portion of the external tank's outer foam. Today and tomorrow engineers will further evaluate worker access to any additional damaged areas by last weekend's hailstorm. If Shuttle managers determine that all repair efforts can be completed at the launch pad, the additional work will not impact Discovery's May 20 launch date. However, if evaluations reveal that significant work must be performed in the Vehicle Assembly Building the launch date could be delayed by one week. Platforms in the VAB provide workers with access to the tank that is not available at the pad. A decision is expected tomorrow afternoon. The hail damage poses no threat to the external tank's structural integrity, but managers are evaluating the likelihood that ice could form inside the divots once the tank is loaded with super-cold propellants and possibly impact the orbiter during ascent. Bruce Buckingham. (1999). **Kennedy Space Center Space Shuttle Status Report** [Online]. Available E-mail: domo@news.ksc.nasa.gov/subscribe shuttle-status [1999, May 12].]

◆ The first shuttle mission of 1999 might not fly next week as planned. NASA could have to roll shuttle Discovery off its launch pad and into a Kennedy Space Center building to repair hail damage to the spaceship's fuel tank. Such a move could delay the May 20 launch for at least a week. However, if the damage is minor, engineers could decide Discovery can be launched safely without the repairs. Discovery was damaged during last weekend's storms when hail pelted the top portion of the tank, which is exposed to the weather. KSC spokesman Bruce Buckingham said KSC workers found about 150 small nicks in the tank's insulation, most of which are less than a half-inch in diameter. Some were as large as 2 inches wide and one-third of an inch thick. The concern is that ice could form inside the holes after the tank is filled with super-cold fuel, then potentially damage the shuttle if it breaks off in flight. Buckingham said most of the holes will be repaired on the launch pad, but about 13 nicks are in a troublesome region that can't be reached. Workers on Wednesday built scaffolding so they could get a better look at the damaged area. Once before, NASA had to roll a shuttle off its launch pad to repair an external tank. The 1995 damage was caused by woodpeckers, which chipped into shuttle Discovery's tank and left many nicks in an area that couldn't be reached from the pad. ["Hail-damaged tank may delay shuttle launch," Florida Today, May 13, 1999, p 1A.]

MAY 13: Space Shuttle Status Report, Thursday, May 13, 1999. STS-96: NASA Shuttle managers decided today to roll Space Shuttle Discovery back to the Vehicle Assembly Building to complete repair work on the hail-damaged external tank foam insulation. After much evaluation, managers determined that necessary repair work could not be performed at Launch Pad 39B due to limited access to damaged areas. Managers expect Discovery's move toward the VAB to begin early Sunday morning, May 16. Current work plans indicate that the foam repairs will take 2 to 3 days, allowing Discovery to roll back to Pad 39B by midweek. Managers expect the Shuttle to be ready for launch no earlier than May 27. Workers will move the STS-93 external tank and booster stack out of VAB high bay 1 Saturday, May 15, to accommodate Shuttle Discovery's return. Once in the VAB, external tank access will be established immediately and workers will inspect the areas that were inaccessible at the pad. At least 35 foam divots have been identified for repair to date. Last weekend, a hailstorm at KSC left about 150 divots in the outer foam of Discovery's external fuel tank. The average diameter of the divots is about 0.5 inches with the largest measuring about 2 inches in diameter. The depth of the dings range from 0.1 to 0.34 inches deep. Shuttle managers' primary concern is that ice could form inside the divots once the tank is loaded with super-cold propellants and then fall off during launch, impacting the orbiter and posing a threat to flight crew safety. Bruce Buckingham. (1999). **Kennedy Space Center Space Shuttle Status Report** [Online]. Available E-mail: domo@news.ksc.nasa.gov/subscribe shuttle-status [1999, May 13].]

◆ NASA decided Thursday to roll Discovery off the launch pad and back to its assembly hangar to repair hail damage from the thunderstorms that battered the Cape last weekend. As a result, the shuttle's May 20 supply flight to the international space station will be delayed at least a week. The rollback – scheduled to begin about midnight Saturday (May 15) – will be the 13th in shuttle history and the second prompted by holes in the external tank's insulation. Last weekend's stormy weather claimed at least one other launch casualty in addition to the shuttle. A military Global Positioning System satellite has been grounded indefinitely after getting rained on Saturday at Cape Canaveral Air Station. After workers left the launch pad because of lightning warnings, water got through tarpaulins covering the satellite's protective clean room atop a Delta II rocket. ["Shuttle Discovery's launch put off for at least 1 week," **The Orlando Sentinel**, May 14, 1999, p A-5.]

MAY 14: Space Shuttle Status Report, Friday, May 14, 1999. STS-96: Shuttle managers decided yesterday to roll Space Shuttle Discovery back to the Vehicle Assembly Building to complete repair work on the hail-damaged external tank foam insulation. Today, KSC managers developed a recovery plan to make Discovery ready for launch no earlier than May 27 if weather conditions cooperate with the current schedule. The plan calls for the STS-93 booster and external tank stack to be removed from VAB high bay 1 on Saturday at 8 a.m. to make room for Discovery's return. The STS-93 stack will temporarily reside at the Mobile Launcher Platform Refurbishment Site until Monday. Shuttle Discovery is slated to begin its 6-hour trip from Pad 39B to the VAB at 4 a.m. Sunday. On Monday morning, the STS-93 stack will transfer to Pad 39B to take advantage of its weather protection system. When Discovery arrives inside the VAB, workers will establish access to the tank's damaged areas and a full inspection of the tank will reveal exactly how much repair work is needed. Preliminary reports indicate that at least 35 foam divots must be repaired over 3 days. Once the necessary repairs are complete, the Space Shuttle could begin moving back to the launch pad as early as Thursday, May 20 at 4 a.m. After Discovery returns to the launch pad, workers must repeat ordnance installation and checks, pressurization of the Shuttle's propellant system, aft compartment close-outs and solid rocket booster close-outs before proceeding with the launch countdown. Bruce Buckingham. (1999). **Kennedy Space Center Space Shuttle Status Report** [Online]. Available E-mail: domo@news.ksc.nasa.gov/subscribe shuttle-status [1999, May 14].]

◆ Due to the recent anomalies involving expendable vehicles not associated with NASA launches, there is a schedule impact on the near-term NASA manifest. NASA is a participant in the failure investigations and is reviewing the readiness of its missions based on an understanding of the failures and the proposed corrective actions. With the exception of the Pegasus/TERRIERS launch at Vandenberg Air Force Base, CA, scheduled for next week, other launch dates are uncertain. Following is a list of upcoming NASA flights with their earliest possible launch timeframes. The launch of TERRIERS for Boston University on a Pegasus rocket from Vandenberg Air Force Base is confirmed for Monday, May 17, during a launch window that extends from 10:05:47 - 10:14:26 p.m. PDT. The drop from the L-1011 aircraft is targeted to occur at 10:10 p.m. PDT. The launch of GOES-L for NASA and NOAA aboard an Atlas IIA rocket from Pad 36-A at Cape Canaveral will occur no earlier than late May following launch of STS-96. The launch of FUSE for NASA and Johns Hopkins University from Pad 17-A at Cape Canaveral will occur no earlier than June 18. This could change to a later date during June. The launch of QuickScat for NASA and the Jet Propulsion Laboratory aboard a Titan II rocket from SLC-4W at Vandenberg Air Force Base will occur no earlier than mid-June. Except for TERRIERS, these dates should be considered for planning purposes until firm dates can be scheduled. ["Launch dates for upcoming NASA spacecraft on expendable vehicles become uncertain," **NASA News Release #36-99**, May 17, 1999.]

◆ President Clinton wants an investigation into a string of launch failures that have cost taxpayers more than $3 billion and jeopardized American's ability to deploy critical spy satellites. Clinton is expected to ask Defense Secretary William Cohen on Monday to probe the failures and examine how future Pentagon satellites should be launched into orbit. The review could include a fresh look at returning at least some defense and intelligence payloads to NASA's shuttle fleet. Between 1985 and 1991 shuttles carried into orbit seven large spy satellites for the Pentagon. Since August, three missions of the Lockheed Martin-built Titan 4 rocket have failed. One mission ended in a fiery explosion over Cape Canaveral. The other two went awry when upper stage rockets malfunctioned. Lost were three military intelligence and communications satellites. In April, a smaller Lockheed Martin Athena rocket failed to deliver a commercial space-imaging satellite into orbit. Boeing's new Delta 3 heavy-lift launcher also has failed twice in the last nine months on commercial missions. The Air Force, Lockheed Martin and Boeing are conducting separate, independent investigations. But some lawmakers and military officials had been seeking to bump the probes to a higher level. ["Clinton wants look into rocket failures," Florida Today, May 15, 1999, p 1A & 2A.]

MAY 16: The Canadian Space Agency's first contribution to the International Space Station (ISS), the Space Station Remote Manipulator System (SSRMS) arrived at KSC May 16 to begin prelaunch processing activities. The SSRMS is the primary means of transferring payloads between the orbiter payload bay and the ISS for assembly. The 56-foot robotic arm has three segments comprising two 12-foot booms jointed by a hinge. Seven joints on the arm allow for highly flexible and precise movement. Cameras on the booms will permit the astronauts maximum visibility for operations and maintenance tasks on the ISS. The SSRMS is scheduled to be launched in July 2000 on STS-100. [**KSC Countdown**, May 20, 1999.]

MAY 17: Space Shuttle Status Report, Monday, May 17, 1999. STS-96: Over the weekend, KSC workers made the necessary flight hardware moves to accommodate this week's repair work on Shuttle Discovery's external tank. On Saturday, the STS-93 solid rocket boosters and external tank were rolled out of VAB high bay 1 and now temporarily reside at the Mobile Launcher Platform refurbishment site east of the VAB and will remain there as long as weather permits. Sunday, before 4 a.m., Shuttle Discovery departed Pad 39B and arrived in VAB high bay 1 at about 10 a.m. Workers then set up access to the areas on the tank that need repair. Today, technicians conducted close-up inspections of the damaged areas, evaluating the exact number of hail-divots and precisely measuring their dimensions. If repair efforts can be completed by Wednesday, Discovery may roll out to Launch Pad 39B as early as Thursday. Thus far, excellent weather conditions have supported a flawless execution of the STS-96 recovery plan. However, current forecasts indicate an increased chance of thunderstorms by Thursday afternoon. Shuttle managers are developing weather protection plans to ensure the safekeeping of all flight hardware involved in these major move operations. Based on the amount of repair work accomplished on Discovery's tank and an updated weather forecast, Shuttle managers will implement a plan tomorrow afternoon that returns Discovery to the launch pad and brings the STS-93 stack back to the VAB. If all goes well, Discovery could be ready to launch no earlier than May 27. Bruce Buckingham. (1999). **Kennedy Space Center Space Shuttle Status Report** [Online]. Available E-mail: domo@news.ksc.nasa.gov/subscribe shuttle-status [1999, May 17].]

◆ The departure of Brig. Gen. Randall Starbuck from his command of the 45th Space Wing at Patrick Air Force Base has been postponed indefinitely. Air Force officials made the announcement Monday but would not give a reason for the delay. The change in plans might mean his replacement, Brig. Gen. Kevin Chilton, no longer is taking over command of the wing, which oversees military and commercial launches from Cape Canaveral Air Station. Chilton, a former NASA space shuttle commander, is deputy manager of operations at Air Force Space Command in Colorado Springs,

Colo. The delay also could mean Starbuck must stay longer to participate in several investigations into recent rocket failures from Cape Canaveral. ["45th Space Wing chief's departure delayed," <u>Florida Today</u>, May 18, 1999, p 1A.]

MAY 18: Space Shuttle Status Report, Tuesday, May 18, 1999. STS-96: Yesterday, technicians completed evaluations on Discovery's hail-damaged external tank foam insulation and began repair efforts. Having much closer access than what is available at the launch pad, inspections in the VAB revealed a total of 648 divots in the tank's outer foam. Managers consider 189 of the divots acceptable to fly without repair. Blending or sanding work is required for 211 hits and 248 divots will be patched with new foam. The current schedule indicates that foam repair efforts will be complete tomorrow. Though repair efforts are going very well, forecasters expect weather conditions at KSC to degrade by Wednesday. The increased chance of rain, wind and lightning have managers implementing a plan that focuses on protecting Space Shuttle flight hardware. Wednesday, the STS-93 solid rocket boosters and external tank will be moved from the Mobile Launcher Platform refurbishment site to Launch Pad 39B to take advantage of available lightning protection. In addition, workers will prepare the STS-99 partial booster stack in VAB high bay 3 for transfer to a recently installed lightning protection system on the east side of the VAB, making room for the STS-93 stack to return later this week. Once the STS-93 stack is removed from Pad 39B, Discovery will roll back out to the launch pad as early as this weekend. The timing of these operations is dependent upon weather. Shuttle managers will meet again tomorrow afternoon to assess the impact that these developments will have on the planned launch date. Bruce Buckingham. (1999). **Kennedy Space Center Space Shuttle Status Report** [Online]. Available E-mail: domo@news.ksc.nasa.gov/subscribe shuttle-status [1999, May 18.]

MAY 19: Space Shuttle Status Report, Wednesday, May 19, 1999. STS-96: Shuttle managers today confirmed May 27 as the launch date for Shuttle mission STS-96. Based on an updated weather forecast and prompt repairs of Discovery's hail-damaged external tank foam insulation, managers decided to transfer Space Shuttle Discovery back to Launch Pad 39B Thursday morning. Shuttle close-out work will commence once Discovery arrives at the pad and will conclude in time for May 27 launch. The launch countdown begins at 9 a.m. on Monday, May 24. Discovery's external tank repairs were completed today and tonight workers will remove access platforms and scaffolding in preparation for Thursday's move to the pad. If weather continues to cooperate, Discovery will begin moving out of VAB high bay 1 about 2 a.m. tomorrow and will be hard down at Pad 39B by midday. The STS-93 stack remains under temporary lightning protection on the east side of the VAB. When VAB high bay 1 becomes available, the STS-93 stack will move inside the VAB to continue processing for its planned July launch. Bruce Buckingham. (1999). **Kennedy Space Center Space Shuttle Status Report** [Online]. Available E-mail: domo@news.ksc.nasa.gov/subscribe shuttle-status [1999, May 19].]

◆   NASA took the unusual position Wednesday of opposing a three-year, $41 billion authorization bill that would give more money that President Clinton had requested. One of the key sticking points for the agency: an amendment by Rep. Dave Weldon, R-Palm Bay, canceling Vice President Al Gore's $75 million pet satellite project, the Triana science mission. The committee eliminated money for the Triana satellite, which would have beamed back live pictures of the sunlit side of Earth for broadcast on the Internet. "We strongly object to the (Science) committee's arbitrary and partisan recommendation to terminate the Triana science mission," NASA Administrator Dan Goldin wrote in a letter. Proposed by Gore and quickly endorsed by NASA, the craft also would be equipped with scientific instruments to take solar, climate, ultraviolet and organic measurements. Weldon said financing the Gore space camera was an insult to former space industry workers in his district who lost their jobs a year ago when NASA said it was facing a $100 million shortfall. In his letter, Goldin warned the authorization bill also could jeopardize development of the $70 billion International Space Station by limiting NASA's ability to continue diverting research dollars to compensate for cost overruns caused by Russia's delays in delivering a key station module. ["Amendment causes NASA to oppose its funding bill," **Florida Today**, May 20, 1999, p 7A.]

MAY 20: Space Shuttle Status Report, Thursday, May 20, 1999. STS-96: Yesterday, technicians completed repair work on more than 460 hail-inflicted divots in time to support Shuttle Discovery's return to Launch Pad 39B this morning. Atop the crawler transporter, Discovery began first motion at about 2 a.m. today and arrived at the pad at about 10 a.m. The STS-93 solid rocket booster and external tank stack moved inside Vehicle Assembly Building high bay 1 at about noon today. With the Rotating Service Structure in place around the Shuttle, launch pad validations are now under way. Workers will complete ordnance connections on Saturday along with pressurization of the orbiter's reaction control system. Routine external tank purge activities occur Sunday. The launch countdown begins Monday at 9 a.m. The STS-96 astronaut crew arrives at KSC late Sunday night to begin final preparation for flight. Bruce Buckingham. (1999). **Kennedy Space Center Space Shuttle Status Report** [Online]. Available E-mail: domo@news.ksc.nasa.gov/subscribe shuttle-status [1999, May 20].]

◆   A new design for the license plate that commemorates the 1986 space shuttle disaster that killed seven astronauts will be unveiled June 25. Officials of the two organizations that benefit from proceeds of Challenger-plate sales hope the redesign and a new marketing plan will boost dwindling sales. The Technological Research and Development Authority and the Astronaut Memorial Foundation each receive $7.50 from each plate sold. Sales of the plate generated $1.2 million for the organizations in 1997, down from $4.2 million during the plate's first year. A total of 600,000 have been sold statewide since the debut in 1987. ["Challenger tag getting new look," **Florida Today**, May 21, 1999, p 1A.]

◆ The former astronaut named the next commander of the 45th Space Wing at Patrick Air Force Base won't be coming to Florida after all, the Air Force announced Thursday. Instead, Brig. Gen. Kevin Chilton will lead the 9th Reconnaissance Wing at Beale Air Force Base near Sacramento, Calif. A new replacement for the 45th's boss, Brig. Gen. Randall Starbuck, has not been chosen. ["Ex-astronaut named to lead Space Wing reassigned," Florida Today, May 21, 1999, p 1A.]

◆ The Air Force estimates it will cost more than $1 million to repair a military navigation satellite that was rained on at Cape Canaveral Air Station. Air Force officials have formed an accident investigation board to determine how the $40 million Global Positioning System IIR satellite got wet during thunderstorms May 8. The satellite was undergoing testing at Launch Complex 17A atop a Delta II rocket when technicians evacuated the pad because of lightning warnings. After workers returned, they found rain had leaked into the satellite's protective clean room on the pad's service tower. The satellite was removed from the rocket last week and returned to its processing facility at the Cape where inspections continue. ["Water damage to satellite put at $1 million," The Orlando Sentinel, May 21, 1999, p A-3.]

MAY 21: Space Shuttle Status Report, Friday, May 21, 1999. STS-96: Launch pad validations are in progress and countdown preparations are under way. Last night, workers began closing out Discovery's aft compartment for flight and expect that work to conclude midday on Monday. Ordnance connections begin today and conclude Saturday. Mating of the orbiter midbody umbilical unit is in work today and Discovery's maneuvering system and reaction control system will be pressurized for flight over the weekend. Routine external tank purge activities occur Sunday. The launch countdown begins Monday at 9 a.m. The STS-96 astronaut crew arrives at KSC Sunday night to begin final preparation for flight. Bruce Buckingham. (1999). **Kennedy Space Center Space Shuttle Status Report** [Online]. Available E-mail: domo@news.ksc.nasa.gov/subscribe shuttle-status [1999, May 21].]

◆ NASA is one of the government's most computer-savvy agencies, yet its crucial information technology systems are vulnerable to hackers, industrial spies and foreign intelligence agents, a new report warns. In a limited test at one NASA field center, government specialists hacked their way into vital space agency computers using simple passwords like "guest" and "newuser," according to the report from the General Accounting Office. NASA information managers supervised the controlled computer "break-in" at the unidentified field center where there are a number of mission-critical systems. The team of experts from the National Security Agency penetrated computers supporting spacecraft command and control as well as a system that processes and distributes scientific data returned from space. NASA officials said they agreed with most of the GAO report. But not all mission-critical systems are vulnerable, they said. And special security steps are taken before launches. NASA plans to follow through on all of the GAO's recommendations for tighter computer security, said NASA's Sarah Keegan. ["NASA's computers vulnerable, report says,"

Florida Today, May 22, 1999, p 1A & 2A. "NASA computers are hacker heaven," The Orlando Sentinel, May 22, 1999, p A-1 & A-4.]

MAY 22: An Air Force Titan 4 rocket made its first successful flight in nine months Saturday, carrying a top-secret spacecraft in an early-morning launch from California. With liftoff at 5:36 a.m., the Titan 4 rocket carried its classified satellite into an orbit that will cross over the Earth's poles. ["Titan 4 launch a success," Florida Today, May 23, 1999, p 1A.]

MAY 24: NASA is set to begin the countdown today for shuttle Discovery's launch Thursday to the International Space Station, which has been vacant since a crew started construction of the outpost six months ago. Discovery's crew arrived at Kennedy Space Center late Sunday, ready for its mission to take 1 ½ tons of supplies and equipment to the facility. After arriving at KSC, the astronauts are to spend the next three days relaxing and reviewing their mission plans while ground crews continue to prepare the ship for liftoff. If launched Thursday, Discovery will reach the outpost Saturday and dock for a five-day visit. ["Discovery countdown begins today," Florida Today, May 24, 1999, p 1A.]

◆ Space Shuttle Status Report, Monday, May 24, 1999. STS-96: The launch countdown for STS-96 began on schedule today at 9 a.m. at the T-43 hour mark. Over the weekend, workers completed ordnance connections and pressurization of Discovery's maneuvering system and reaction control system. Yesterday morning, aft compartment close-outs concluded and aft confidence checks are in work today. Mating of the orbiter midbody umbilical unit is also complete. This morning, pyrotechnic initiator controller testing began. At about 11 p.m. yesterday, the STS-96 flight crew arrived at KSC's Shuttle Landing Facility. In the three days prior to launch they will undergo routine medical exams and participate in standard orbiter and payload familiarization activities. The commander and pilot will also take practice flights in the Shuttle Training Aircraft and T-38 jets this week. Weather forecasters currently indicate a 30 percent chance that weather could prohibit Thursday's launch attempt. The forecast calls for scattered clouds with bases at 15,000 feet; visibility at 7 miles with a chance of fog limiting to 3 miles; westerly winds at 7 knots peaking to 10 knots; temperature at 72 degrees F; relative humidity at 93 percent and dew point at 70 degrees F. The primary weather issues are expected to be the possibility of fog and low cloud ceilings. The forecast improves to a 20 percent chance of violation on Friday and Saturday. Bruce Buckingham. (1999). **Kennedy Space Center Space Shuttle Status Report** [Online]. Available E-mail: domo@news.ksc.nasa.gov/subscribe shuttle-status [1999, May 24].]

◆ For the first time in six months, countdown clocks started rolling Monday toward a shuttle launch at Kennedy Space Center. Seven astronauts are set to lift off aboard Discovery between 6:48 and 6:57 a.m. Thursday on a 10-day supply mission to the international space station. The six-month gap between missions – created mainly by delays in the station program and problems with an X-ray telescope scheduled to launch aboard Columbia – is the longest since the shuttle's September 1988 return to flight after the January 1986 Challenger accident. In an effort to do things right, mission managers at KSC have conducted extensive drills and training since the last shuttle flight in December to keep everyone sharp. Discovery's mission is the second of 36 shuttle flights planned to assemble the space station. Astronauts will drop off 3,600 pounds of supplies and spare parts at the station, do maintenance work and conduct a spacewalk to erect a crane on the outpost's hull. ["Countdown for shuttle launch," The Orlando Sentinel, May 25, 1999, p A-3.]

◆   A top Russian space official left jail Monday charged with attacking two emergency medical workers. Vladimir Lobachev, 61, of Moscow is in Florida as a visiting dignitary for Thursday's launch of the space shuttle Discovery, a NASA spokesman said. Lobachev posted $1,000 bail on two charges of battery on an emergency medical worker and left the jail shortly after 2 p.m., records show. NASA spokesman Bruce Buckingham said Lobachev has no active role in the upcoming shuttle mission. ["Police: Russian space chief attacked EMTs," The Orlando Sentinel, May 25, 1999, p D-1 & D-5.]

MAY 25: Space Shuttle Status Report, Tuesday, May 25, 1999. STS-96: Preparation for Discovery's Thursday launch continues on schedule. Last night, the launch team activated the orbiter's navigational systems and completed testing efforts. Pyrotechnic initiator controller testing began at about 2 a.m. today and loading of cryogenic reactants into Discovery's fuel cell storage tanks commenced at 6:30 a.m. Workers will demate the orbiter midbody umbilical unit today at 1:30 p.m. Tonight, technicians will install the mission specialist seats inside the crew compartment and begin cable routing verifications. The flight crew wakes-up today at 5:30 p.m. and will participate in standard preflight orbiter and payload system briefings this evening. The commander and pilot will begin practice flights in the Shuttle Training Aircraft at about 10 p.m. Weather forecasters currently indicate a 20 percent chance that weather could prohibit Thursday's launch attempt. The forecast calls for scattered clouds with bases at 20,000 feet; visibility at 7 miles with a chance of fog limiting to 3 miles; southwesterly winds at 7 knots peaking to 10 knots; temperature at 72 degrees F; relative humidity at 93 percent and dew point at 70 degrees F. The primary weather issues are expected to be the possibility of fog and low cloud ceilings. The 20 percent chance of weather violation continues through Friday and Saturday. Bruce Buckingham. (1999). Kennedy Space Center Space Shuttle Status Report [Online]. Available E-mail: domo@news.ksc.nasa.gov/subscribe shuttle-status [1999, May 25].]

◆   NASA took a stand Tuesday on the idea of relying on automatic destruct systems – rather than human engineers – to deliberately destroy an errant space shuttle with men and women on board. Simply stated, agency officials say the notion is unacceptable. And both commercial launch companies and the Department of Defense aren't all that enthusiastic about the concept either, especially when it comes to multimillion satellite-delivery missions or ballistic missile test flights. That was the message aerospace industry leaders delivered to a National Research Council committee conducting an independent study of the way the Air Force protects the public from runaway or exploding rockets. In the second of two public hearings held in Cocoa Beach, the committee fielded testimony on proposed changes to a "range safety" system that enables engineers to deliberately destroy an errant launcher before it can threaten populated areas. The proposals all are aimed at cutting the cost of protecting the public. One involves the idea of relying solely on automatic destruct systems rather than human engineers to obliterate space launchers that careen out of control. "For those of us in the human space flight program, an automatic destruct system is pretty much unacceptable," said former astronaut Loren Shriver, who now is deputy director of launch and payload processing at NASA's Kennedy Space Center. In the event of a serious in-flight failure on a shuttle mission, an automatic destruct system would do just that – it automatically would blow up the ship with its crew onboard. A range safety engineer, however, could make a judgment call, giving an astronaut crew every possible opportunity to maneuver a shuttle out of harm's way and on to an emergency landing site. "Our preference is to have man in the loop," Shriver said. "That part of it is not rocket science at all. It's just common sense." Commercial launch companies and the U.S. Navy prefer to keep human engineers involved in the process for similar reasons. ["NASA: Humans should control destruct switch," Florida Today, May 26, 1999, p 6A.]

MAY 26: Space Shuttle Status Report, Wednesday, May 26, 1999. STS-96: Preparations for launch of Space Shuttle Discovery on Thursday continue as scheduled today. The Rotating Service Structure was retracted away from the vehicle and placed in launch configuration early this afternoon. Earlier today, the orbiter's inertial measurement units were activated and the flight crew's equipment was stowed in the orbiter. Currently, final walkdowns of the pad are being completed in preparation for loading of the 500,000 gallons of cryogenic propellants into the Shuttle's external tank. Loading operations will begin at about 9:30 tonight. The crew will begin final preparations for launch when they awake at about 6:30 p.m. today. They are scheduled to begin donning their launch and entry suits at 2:19 a.m. tomorrow and depart for the launch pad at about 3 a.m. Weather forecasters currently indicate a 20 percent chance that weather could prohibit Thursday's launch attempt. The forecast calls for scattered clouds with bases at 20,000 feet; visibility at 7 miles with a slight chance of early morning fog; southwesterly winds at 7 knots peaking to 10 knots; temperature at 72 degrees F; relative humidity at 93 percent and dew point at 70 degrees F. The crew for mission STS-96 are: Commander Kent Rominger, Pilot Rick Husband, and Mission Specialists Tamara Jernigan, Ellen Ochoa, Daniel Barry, Julie Payette, and Valerie Tokarev. Bruce Buckingham. (1999). Kennedy Space Center Space Shuttle Status Report [Online]. Available E-mail: domo@news.ksc.nasa.gov/subscribe shuttle-status [1999, May 26].]

MAY 27: Space Shuttle Status Report, Thursday, May 27, 1999. STS-96: Space Shuttle Discovery and the seven-member STS-96 flight crew lifted-off on time today at 6:49:42

a.m. The launch team worked no significant technical issues throughout the day's countdown activities. Today's launch marks the beginning of Shuttle Discovery's 26th flight and the 94th Shuttle flight in history. Discovery and crew will rendezvous and dock with the International Space Station on Flight Day 3. The Shuttle and ISS will remain docked for six days while the crew transfers some 4,000 pounds of equipment, tools, computers and clothes to the ISS for future use by downstream construction crews. Solid Rocket Booster recovery operations are under way, off the eastern coast of Florida NASA's recovery ships Freedom Star and Liberty Star are expected to return to Port Canaveral with the boosters in tow tomorrow. Bruce Buckingham. (1999). **Kennedy Space Center Space Shuttle Status Report** [Online]. Available E-mail: domo@news.ksc.nasa.gov/subscribe shuttle-status [1999, May 27].]

◆   NASA honored the late CNN space correspondent John Holliman on Thursday by bestowing his name on the Kennedy Space Center press auditorium. With Holliman's widow and son nearby, NASA chief Dan Goldin dedicated the John Holliman Auditorium and praised the former space reporter for his enthusiasm. "He enriched the lives of everyone who knew him," Goldin said during a ceremony that followed shuttle Discovery's launch. "The naming of this auditorium will serve as a reminder of how John touched all our lives. John was a true fan of the space program. "He conveyed that sense of excitement to his audience. John didn't simply report on space issues, he conveyed the deep passion he had for space issues," Holliman was killed in September in a car accident near his suburban Atlanta home. He joined CNN in 1980 as part of the network's original reporting team and gained notoriety on the CNN team covering the Persian Gulf War from Baghdad. ["NASA honors late CNN reporter," <u>Florida Today</u>, May 28, 1999, p 2A.]

◆   An Air Force investigation could conspire to delay the planned launch next month of shuttle Columbia, officials said Thursday. Columbia and five astronauts are tentatively scheduled to lift off July 22 on a mission to deploy NASA's Chandra X-Ray Astrophysics Facility, a $1.5 billion astronomical observatory. That date, however, is threatened by an Air Force investigation into the failed April 9 launch of a Titan 4 rocket and a $250 million missile warning satellite. A two-stage, solid-fueled booster did not separate during that mission, and the missile warning satellite was delivered to a useless orbit as a result. The same type of booster is to be used to deploy the Chandra observatory. NASA officials said the mission won't be launched until the Air Force completes its investigation into the April 9 failure. Unclear is exactly when the probe will be complete. Senior NASA managers will meet Tuesday to decide whether to continue with preparations for a July 22 launch. ["Air Force probe may delay next launch," <u>Florida Today</u>, May 28, 1999, p 4A.]

MAY 28: Space Shuttle Status Report, Friday, May 28, 1999. STS-96: Shuttle Discovery's systems are performing very well on orbit. Post launch inspections of Launch Pad 39B revealed minimal pad damage. Inspectors did collect several samples of a residue found on various portions of the Fixed Service Structure. Managers expect laboratory evaluations to confirm that the residue resulted from a reaction between solid rocket booster (SRB) exhaust and the protective coating applied to the exterior of the FSS. The SRB recovery ships arrived at Hangar AF this afternoon with both SRBs in tow. Preliminary reports indicate that the boosters are in good condition. STS-93: Technicians have completed Columbia's closeouts in preparation for next week's OPF roll out. Weight and center of gravity testing began today and managers plan to transfer Columbia to the Vehicle Assembly Building on Wednesday at about 10 a.m. The Chandra payload resides in KSC's Vertical Processing Facility awaiting arrival of the Inertial Upper Stage. The IUS is scheduled to arrive in the VPF Tuesday for mating operations with Chandra on Wednesday. Following electrical and mechanical testing, the spacecraft is slated for transfer to Launch Pad 39B on June 21. STS-99: Fabrication and installation of tubing for Endeavour's freon coolant loop No. 2 continues. Modifications on the main engine dome heat shields and wireless video installation efforts continue as well. STS-101: Installation of Atlantis' power control assembly No. 1 is complete. Wireless video modifications continue. Mass spectrometer leak checks on the orbiter's fuel cell power plant continue. TACAN antenna installation is in progress. Bruce Buckingham. (1999). **Kennedy Space Center Space Shuttle Status Report** [Online]. Available E-mail: domo@news.ksc.nasa.gov/subscribe shuttle-status [1999, May 28].]

MAY 29: A multinational crew made an early-morning rendezvous with NASA's International Space Station, pulling off the first-ever space shuttle docking at the new $60 billion outpost. With four Americans, a Russian and a Canadian aboard, shuttle Discovery linked up with the seven-story station at 12:24 a.m. today as both craft cruised along at 17,500 mph. The shuttle and the station – made up of a Russian space tug and a U.S. docking module – were passing 230 miles above a Russian ground station in Central Asia at the time. "The history of this moment shouldn't be lost on us," said Frank Culbertson, NASA's deputy director for space station operations. Shuttles linked with the Mir space station nine times between 1995 and 1998 as the United States and Russia prepared to build the new outpost. Not since 1973, however, has an American spaceship docked with a U.S. space station. In that case, three astronauts docked an Apollo spacecraft at Skylab – the nation's first space station. ["Shuttle docks at new station for first time," <u>Florida Today</u>, May 29, 1999, p 1A.]

DURING MAY: In addition to NASA's basic operations budget for Kennedy Space Center, NASA is FY-98 had obligations for $488 million in contracts with industry, academia, and other organizations statewide. The funds paid for products and services within 21 of Florida's 23 congressional districts, including $26 million to academic institutions. ["NASA pumps $488 million into Florida economy for FY-1998," **The Brevard Technical Journal**, May 99, p 3.]

◆ Kennedy Space Center and a consortium of other federal agencies are leading the way to develop faster, more economic clean-up methods for numerous contaminated groundwater sites across the country, including an area at Launch Complex 34, Cape

Canaveral Air Station. ["KSC, consortium to study cleanup of groundwater," **The Brevard Technical Journal**, May 99, p 6.]

JUNE

JUNE 1: Space Shuttle Status Report, Tuesday, June 1, 1999. STS-96: Space Shuttle Discovery's onboard systems continue to perform well on orbit. Discovery has been docked with the International Space Station for four days and the crew has completed more than 30 percent of the supply transfer activities. Preliminary assessments of the solid rocket boosters reveal no significant damage. More detailed booster inspections are in work at Hangar AF today. Launch pad wash down efforts are complete and inspections revealed minimal damage. Evaluation of the SRB exhaust residue continues. STS-93: Columbia's weight and center of gravity testing is complete. Today workers are placing Columbia on the orbiter transporter system in preparation for tomorrow's transfer to the Vehicle Assembly Building. Managers plan to move Columbia to the VAB Wednesday at about 10 a.m. Chandra's Inertial Upper Stage arrived at KSC's Vertical Processing Facility this morning and mating to the X-ray observatory begins tomorrow. Following electrical and mechanical testing, the spacecraft is slated for transfer to Launch Pad 39B on June 21. STS-99: Workers completed installation of Endeavour's window No. 6. Shuttle main engine installation efforts are under way and modifications to the main engine dome heat shields continue. Wireless video installation efforts proceed on schedule. Modifications to the freon coolant loop No. 2 isolation valve are in work. Checks on the auxiliary power unit exhaust duct and power reactant storage and distribution system are also in progress. Installation of Endeavour's right hand orbital maneuvering system pod begins later today. STS-101: Testing on Atlantis' flash evaporator system controller is complete. Wireless video modifications continue. Mass spectrometer leak checks on the orbiter's fuel cell power plant continue. Ammonia system leak and functional testing begins later today. Bruce Buckingham. (1999). **Kennedy Space Center Space Shuttle Status Report** [Online]. Available E-mail: domo@news.ksc.nasa.gov/subscribe shuttle-status [1999, June 1].]

◆ Payload Status Report – Chandra X-Ray Observatory, Tuesday, June 1, 1999. The Inertial Upper Stage (IUS) booster for the Chandra X-ray Observatory arrived at the Kennedy Space Center's Vertical Processing Facility (VPF) at 6 a.m., Tuesday, June 1. This evening it will be lifted from it's enclosed transporter and installed into the VPF west test cell. On Wednesday evening, June 2, the Chandra X-ray Observatory will be hoisted from its location in front of the east test cell and mated onto the two-stage IUS. On Thursday, June 3, the electrical connections between Chandra and the IUS will be established and the Cargo Integrated Test Equipment (CITE), an orbiter avionics simulator, will be connected to the payload stack. On Friday, June 4, the two-day interface verification test (IVT) is to be performed which validates the IUS/Chandra connections and checks the orbiter avionics interfaces. On Monday, June 7, the end-to-end test (ETE) is to be conducted. This will verify the communications path to Chandra, commanding it as if it were in space. Participating will be Chandra's operations control center located in Cambridge, MA; Mission Control at the Johnson Space Center in Houston; and the communications assets of both the Deep Space

Network and the Tracking and Data Relay Satellite (TDRS) system. During Chandra's recent dwell period in the VPF, while awaiting the arrival of the IUS, the command paths to be used during these upcoming activities were able to be established and thoroughly tested. Upon successfully completing these tests, the IUS/Chandra payload separation ordnance are to be installed and payload closeouts will be performed in preparation for making the transition to Launch Pad 39-B. The Chandra/IUS combination will be hoisted into the payload canister on June 18 and transported to the pad payload changeout room on June 21. Chandra is currently planned to be installed into Columbia's payload bay on June 25. At this time, the STS-93 launch is targeted for July 22, however, this date is under review. Bruce Buckingham. (1999). **Payload Status Report, Chandra X-Ray Observatory** [Online]. Available E-mail: domo@news.ksc.nasa.gov/subscribe shuttle-status [1999, June 1].]

JUNE 2: Space Shuttle Status Report, Wednesday, June 2, 1999. STS-96: Flight controllers report that Discovery's onboard systems are in excellent health. Flight crew efforts to transfer about two tons of supplies to the International Space Station conclude today. Tomorrow, following a reboost of the ISS, the two spacecraft will undock. Solid rocket booster inspections and segment disassembly preparations are ongoing. All booster components appear to be in good condition following a normal splash down. Tomorrow, SRB aft exit cone disconnects begin and Thursday aft skirt removal begins. The solid rocket motor nozzles will be removed Friday. STS-93: Columbia was transferred from Orbiter Processing Facility bay 1 to the VAB today at about 10:30 a.m. The orbiter will be lifted into VAB high bay 1 today to be mated to the external tank and solid rocket booster stack on Mobile Launcher Platform 1. The Chandra X-ray Observatory and Inertial Upper Stage are being mated in the Vertical Processing Facility today. Later this week electrical and mechanical testing will follow. The spacecraft remains on schedule for transfer to Launch Pad 39B on June 21. STS-99: Modifications to the main engine dome heat shields continue. Wireless video installation efforts proceed on schedule. Modifications to the freon coolant loop No. 2 isolation valve are in work. Checks on the auxiliary power unit exhaust duct and power reactant storage and distribution system are also in progress. Installation of Endeavour's right hand orbital maneuvering system pod begins this week. STS-101: Testing on Atlantis' flash evaporator system controller is complete. Wireless video modifications continue. Mass spectrometer leak checks on the orbiter's fuel cell power plant continue. Ammonia system leak and functional testing is scheduled this week. Bruce Buckingham. (1999). **Kennedy Space Center Space Shuttle Status Report** [Online]. Available E-mail: domo@news.ksc.nasa.gov/subscribe shuttle-status [1999, June 2].]

JUNE 3: Space Shuttle Status Report, Thursday, June 3, 1999. STS-96: Shuttle Discovery is in its final day of docked operations with the International Space Station. Equipment transfer activities concluded early this morning and efforts to boost the space station's orbital position were successfully completed at about 6 a.m. Discovery's systems continue to perform very well as the flight crew prepares to undock from the ISS later today. Solid rocket booster inspections and segment disassembly are in work.

Removal of the solid rocket motor nozzles is ongoing and nozzle shipment to Utah is slated for Monday. The preliminary weather forecast for Discovery's KSC landing on Sunday calls for scattered clouds at 3,000 feet and 10,000 feet and broken at 20,000 feet; visibility at 7 miles; winds east north east at 8 knots peaking to 14 knots; temperature at 75 degrees F; humidity at 91 percent and the possibility of showers within 30 nautical miles of the runway. Forecasters will monitor the development of a surface low over the Bahamas and its impact on landing day weather conditions. STS-93: Columbia arrived in the VAB yesterday and soft mate of the orbiter to the external tank was completed last night. Electrical and mechanical connections are in work today and the Shuttle Interface Test begins Friday at midnight. Electrical and mechanical testing of the mated Chandra X-ray Observatory and Inertial Upper Stage is scheduled this week. The spacecraft remains on schedule for transfer to Launch Pad 39B on June 21. STS-99: Auxiliary power unit exhaust duct leak checks are complete. Testing of Endeavour's radar altimeter has also concluded. Modifications to the main engine dome heat shields continue. Wireless video installation efforts proceed on schedule. Modifications to the freon coolant loop No. 2 isolation valve are in work. Power reactant storage and distribution system are in progress. Installation of Endeavour's right hand orbital maneuvering system pod begins early tomorrow and Shuttle main engine installation begins Monday. STS-101: Atlantis' wireless video modifications continue. Mass spectrometer leak checks on the orbiter's fuel cell power plant continue. Ammonia system leak and functional testing is scheduled this week. Bruce Buckingham. (1999). **Kennedy Space Center Space Shuttle Status Report** [Online]. Available E-mail: domo@news.ksc.nasa.gov/subscribe shuttle-status [1999, June 3].]

◆ After successfully attaching the two cranes to the outside of the ISS, the crew has been transferring two tons of supplies from Discovery to the ISS as well as replacing some equipment. Discovery is scheduled to return to Earth and land at KSC's Shuttle Landing Facility on Sunday, June 6, at 1:59 a.m. EDT. ["STS-96 at the International Space Station," **KSC Countdown**, June 3, 1999.]

◆ STS-96/Discovery Landing Weather Forecast, Thursday, June 3, 1999. Synopsis: A surface and upper level high pressure ridge across northern Florida will cause an easterly flow across central Florida on Sunday morning. Forecast development of a surface low over the Bahamas will be key to the strength of the easterly wind and extent of showers off shore at landing time. Kennedy Space Center Shuttle Landing Facility: Valid: 0200 EDT, Sunday, June 6; Clouds: 3,000 scattered; 10,000 scattered; 20,000 broken; Visibility: 7 miles; Wind: ENE at 8 knots, peaks to 14 knots; Temperature: 75; Dewpoint: 73 Relative Humidity: 91; Precipitation: Rain showers within 30 nautical miles over the water; Edwards Air Force Base: Valid: 0211 PDT; Clouds: 5,000 few; 25.000 scattered; Visibility: 7 miles; Wind: SW 6 knots, peaks to 10 knots; Forecast prepared by NOAA/NWS Space Flight Meteorology Group. Bruce Buckingham. (1999). **STS-96/Discovery Landing Weather Forecast** [Online]. Available E-mail: domo@news.ksc.nasa.gov/subscribe shuttle-status [1999, June 3].]

◆ Shuttle Discovery's crew bid farewell to the international space station Thursday after successfully wrapping up NASA's second mission to build and outfit the orbital outpost. Pilot Rick Husband ended the astronauts' five-day stay at the station at 6:39 p.m., popping a series of hooks and latches that held the shuttle to the station's seven-story stack of modules. Springs in the docking mechanism pushed Discovery several feet away. Then, Husband fired the shuttle's steering jets to inch Discovery back to a distance of 450 feet. From there, the shuttle flew 2 ½ loops around the station while the crew took photos and videotaped the outpost. During the five days the shuttle was docked to the station, Discovery's seven astronauts unloaded more than 4,200 pounds of cargo for use by the outpost's first full-time inhabitants. After Thursday's undocking, the crew wrapped up a few light chores before taking the rest of the night off. They have one major piece of business remaining before heading home to Kennedy Space Center early Sunday. At about 3 a.m. Saturday, the astronauts are scheduled to deploy the Starshine satellite, a beachball-sized globe covered by 878 aluminum mirrors. ["Discovery leaves space station stocked for future crew," The Orlando Sentinel, June 4, 1999, p A-1 & A-11.]

JUNE 4: Space Shuttle Status Report, Friday, June 4, 1999. STS-96: Following Discovery's successful undocking from the International Space Station yesterday, the flight crew will today begin preparations for the seventeenth consecutive KSC Shuttle landing early Sunday morning. Discovery has two landing opportunities at KSC on Sunday. The first opportunity requires a 12:54 a.m. deorbit burn and touchdown on Runway 15 at 2:03 a.m. The second opportunity calls for a deorbit burn at 2:30 a.m. and touchdown at 3:38 a.m. Edward Air Force Base will not be called to support as an alternate landing site on Sunday. Weather officials expect generally favorable weather for a Shuttle landing at KSC on Sunday. Preliminary reports call for scattered clouds at 3,000 feet and 10,000 feet and broken at 20,000 feet; visibility at 7 miles; easterly winds at 6 knots peaking to 12 knots; temperature at 75 degrees F; humidity at 91 percent and the chance of showers within 30 nautical miles of the runway. Forecasters continue to monitor the development of a surface low over the Bahamas and its impact on landing day weather conditions. STS-93: Columbia is hardmated to the external tank and solid rocket booster stack in VAB high bay 1 today. Tomorrow, the Shuttle Interface Test will conclude and over the weekend workers will prepare for Monday's Shuttle roll out to Launch Pad 39B. The Chandra X-ray Observatory and Inertial Upper Stage are undergoing interface verification testing today. STS-99: Modifications to the main engine dome heat shields continue. Wireless video installation efforts proceed on schedule. Modifications to the freon coolant loop No. 2 isolation valve are in work. Tests on the power reactant storage and distribution system are in progress. Installation of Endeavour's right hand orbital maneuvering system pod begins early tomorrow and Shuttle main engine installation begins Monday. STS-101: Atlantis' wireless video modifications continue. Mass spectrometer leak checks on the orbiter's fuel cell power plant continue. Ammonia system leak and functional testing is scheduled this week. Bruce Buckingham. (1999). **Kennedy Space Center Space**

Shuttle Status Report [Online]. Available E-mail: domo@news.ksc.nasa.gov/subscribe shuttle-status [1999, June 4].]

◆ Discovery's astronauts will attempt a rare, middle-of-the-night landing Sunday, capping a 10-day trip to the International Space Station. With seven astronauts aboard, Discovery is scheduled to make the 11th night landing in 94 shuttle flights, touching down at Kennedy Space Center at 2:03 a.m. NASA mission managers, however, will be keeping close tabs on the weather. Forecasters are calling for a chance of rain within 30 miles of KSC. NASA flight rules would prohibit a landing attempt if that happens. Discovery and its crew will have one additional chance to land at KSC on Sunday. The second will come at 3:38 a.m. Launched May 27, the shuttle has enough fuel to remain in space until Tuesday. ["Discovery set to land while most sleeping," Florida Today, June 5, 1999, p 1A.]

◆ An $800 million military communications satellite was hurled into a useless orbit April 30 when the upper stage of a Titan 4 rocket misfired because of faulty computer software, the Air Force said Friday. The mishap was the sixth in a series of botched launches to sting the U.S. space industry during a nine-month period dating back to August 1998. It was also the third consecutive failure for the Air Force's Titan rocket program, which since has tallied a successful launch from Vandenberg Air Force Base in California. Launched from Cape Canaveral Air Station, the April 30 mission went awry after a Titan 4 rocket dropped off a Centaur upper stage motor and a Milstar 2 satellite in low Earth orbit. Designed to propel the satellite to an orbit 22,300 miles above Earth, the upper stage "began to fire excessively" shortly after it separated from the Titan 4, the Air Force said. All fuel in the upper stage was guzzled prematurely, and the satellite was left in an egg-shaped orbit with a high point of 3,100 miles. The satellite has been declared a total loss. ["Bad software doomed Titan, Air Force says," Florida Today, June 5, 1999, p 2A.]

JUNE 6: Space Shuttle Status Report, Sunday, June 6, 1999. STS-96: Space Shuttle Discovery and the STS-96 flight crew made a safe and successful landing on KSC runway 15. This marks the 11th night landing in Shuttle history and the 18th consecutive landing at KSC's Shuttle Landing Facility.

	MET	EDT
Main Gear Touchdown	9:19:13:1	2:02:43 a.m.
Nose Gear Touchdown	9:19:13:16	2:02:58 a.m.
Wheels Stop	9:19:13:57	2:03:39 a.m.

Following routine safing of the vehicle and preliminary inspections, Discovery will be towed to Orbiter Processing Facility bay 1. Inspections of the orbiter will follow in the OPF along with routine post flight deservicing of the orbiter's main propulsion system, hydraulic power system and other major systems. Discovery's next flight is on mission STS-103 slated for October 14. Bruce Buckingham. (1999). **Kennedy Space Center Space Shuttle Status Report** [Online]. Available E-mail: domo@news.ksc.nasa.gov/subscribe shuttle-status [1999, June 6].]

◆   Discovery's astronauts are back on the ground, but plans for NASA's next shuttle mission, on Columbia and a $1.5 billion astronomy observatory are up in the air. In fact, an Air Force accident investigation, coupled with a planned shuttle overhaul and three back-to-back missions this year, could postpone Columbia's long-delayed flight another year or more. Potential cost to taxpayers is about $8 million a month to store an observatory that otherwise would be giving astronomers an unprecedented view of mysterious black holes and invisible clouds of gas swirling among the stars. "We may have a real problem here," NASA Administrator Daniel Goldin said Sunday after Discovery capped a 10-day space station supply mission with a middle-of-the-night landing at Kennedy Space Center. "But we will not launch (Columbia) until it's safe. And if we have to wait a year, we have to wait a year." Columbia's payload -- an X-ray observatory in the same class as the Hubble Space Telescope – is to be popped out of the shuttle's cargo bay and then propelled by a $70 million upper stage into an egg-shaped orbit that stretches a third of the way to the moon. An identical motor, however, failed during an April 9 Air Force Titan 4 rocket mission, stranding a $250 million missile warning satellite into a useless orbit. Air Force investigators say the two-stage, solid-fueled motor failed to separate in space, sending the Pentagon spacecraft into a wild tumble. Goldin said Columbia will remain grounded until the Air Force pinpoints the cause of the failure and NASA is certain it won't recur on the $2.5 billion telescope mission. Air force investigators are scheduled to meet with NASA officials June 18 to update the agency on the ongoing probe into the ill-fated April 9 flight. ["Discovery lands safely in dark," **Florida Today**, June 7, 1999, p 1B & 2B.]

JUNE 7: Payload Status Report, Chandra X-Ray Observatory, Monday, June 7, 1999. The end-to-end test of the Chandra X-ray Observatory began this morning at 7 a.m. as scheduled. This 24-hour test will verify the communications path to Chandra, commanding it as if it were in space. Participating are Chandra's Operations Control Center located in Cambridge, MA, Mission Control at the Johnson Space Center in Houston, the communications assets of both the Deep Space Network and the Tracking and Data Relay Satellite (TDRS) system.   On Saturday, June 4, the two-day interface verification test was successfully completed with no issues or concerns. This test validated the Chandra/Inertial Upper Stage (IUS) connections and checked the orbiter avionics interfaces. Upon successfully completing the end-to-end test, the Chandra/IUS payload separation ordnance will be installed and payload closeouts will be performed in preparation for making the transition to Launch Pad 39-B. The Chandra/IUS combination will be transferred into the transportation canister on June 18. The payload will now be moved to the pad payload changeout room on June 19 and installed into Columbia's payload bay on June 24. The IUS, a two-stage solid propellant booster, arrived at the Kennedy Space Center's Vertical Processing Facility

(VPF) on Tuesday, June 1. On Wednesday, June 2, the Chandra X-ray Observatory was mated to the IUS. At this time, the STS-93 launch is targeted for July 22, however, this date is under review. Bruce Buckingham. (1999). **Payload Status Report, Chandra X-Ray Observatory** [Online]. Available E-mail: domo@news.ksc.nasa.gov/subscribe shuttle-status [1999, June 7].]

JUNE 8: Space Shuttle Status Report, Tuesday, June 8, 1999. STS-93: Today, Shuttle managers approved a plan that will have Columbia ready for launch no earlier than July 20. As always, an official launch date will be announced at the STS-93 Flight Readiness Review scheduled for July 8. Space Shuttle Columbia rolled out of the Vehicle Assembly building at about 2 a.m. yesterday and was hard down at Launch Pad 39B at about 10 a.m. Launch pad validations are in work through this evening. Auxiliary power unit hot fire testing and main engine tests are scheduled this week. The Chandra/Inertial Upper Stage end-to-end test was successfully completed today in the Vertical Processing Facility. The Chandra/IUS combination will be placed into the transportation canister on June 18. The payload will be moved to the launch pad on June 19 and installed inside Columbia's cargo bay on June 24. STS-99: Checks on Endeavour's power reactant storage and distribution system are complete. Interface verification tests of the recently installed right hand orbital maneuvering system pod are also complete. Shuttle main engine installation is scheduled to begin tomorrow. Modifications to the main engine dome heat shields continue. Wireless video installation efforts proceed on schedule. Modifications to the freon coolant loop No. 2 isolation valve are in work. STS-103: After Shuttle Discovery's Sunday landing at KSC, workers conducted routine inspections of the orbiter's thermal protection system and landing gear. TPS inspections revealed 160 debris hits to the orbiter's lower surface with 66 hits measuring 1-inch or greater. The orbiter's tires are reported to be in excellent condition following a landing with notable cross winds. The drag chute functioned normally, having flown for the first time with upgraded inconel shear pins on the drag chute door. Draining of the power reactant storage and distribution system is complete and orbiter access installation is in work in Orbiter Processing Facility bay 1. Preparations are under way to open the payload bay doors. STS-101: Replacement of Atlantis' flash evaporator system controller is complete. Wireless video modifications continue. Mass spectrometer leak checks on the orbiter's fuel cell power plant continue. Ammonia system leak and functional testing proceeds this week. Bruce Buckingham. (1999). **Kennedy Space Center Space Shuttle Status Report** [Online]. Available E-mail: domo@news.ksc.nasa.gov/subscribe shuttle-status [1999, June 8].]

◆ The Boeing Co. will make a second attempt today to launch a Delta 2 rocket on a satellite-delivery mission, but stormy weather once again could force a flight delay. The 12-story Delta 2 and its payload – a quartet of cellular telephone satellites – are scheduled to lift off from complex 17 at Cape Canaveral Air Station between 10:04 and 10:07 a.m. But forecasters say there is a 40 percent chance thick clouds or thunderstorms could ground the Delta 2 for a second consecutive day. An initial liftoff attempt Tuesday was scrubbed when thick clouds moved within 10 miles of the launch pad late in the countdown. A rocket flying through such clouds could trigger lightning strikes, potentially destroying the $48 million launcher and the four satellites, each of which are valued at about $13 million. ["Weather may again delay launch of Delta 2 rocket," Florida Today, June 9, 1999, p 4A.]

JUNE 10: Space Shuttle Status Report, Thursday, June 10, 1999. STS-93: Launch pad validations are complete at Pad 39B. Columbia's auxiliary power units (APU) No. 1 and No. 3 have completed hotfire testing and APU No. 2 will follow next week. Main engine flight readiness testing concluded yesterday. Planned cycling of the orbiter's payload bay doors also concluded yesterday. Preparations are in work for the Helium Signature Test scheduled through tomorrow. Columbia's hypergolic loading begins Monday. The Chandra/IUS combination will be placed into the transportation canister on June 18. The payload will be moved to the launch pad on June 19 and installed inside Columbia's cargo bay on June 24. STS-99: Installation of Endeavour's main engine No. 1 is complete and installation of engines No. 2 and No. 3 continue this week. Modifications to the main engine dome heat shields continue. Wireless video installation efforts continue on schedule. Modifications to the final section of line on freon coolant loop No. 2 are in work. STS-103: Workers have completed orbiter access installation and the payload bay doors were opened yesterday. Preparations are in work for Monday's scheduled payload removal. Preparations are also under way for Discovery's hypergolic deservicing set to begin next week. Routine post flight inspections of the orbiter continue. STS-101: Leak and functional testing of Atlantis' ammonia system is complete. Wireless video modifications continue. Mass spectrometer leak checks on the orbiter's fuel cell power plant continue. UHF system checks are under way. Bruce Buckingham. (1999). **Kennedy Space Center Space Shuttle Status Report** [Online]. Available E-mail: domo@news.ksc.nasa.gov/subscribe shuttle-status [1999, June 10].]

◆ NASA's Far Ultraviolet Spectroscopic Explorer (FUSE) satellite, developed by The Johns Hopkins University under contract to Goddard Space Flight Center, Greenbelt, Md., is scheduled to be launched aboard a Boeing Delta II rocket June 23 at Launch Pad 17A, Cape Canaveral Air Station. FUSE will investigate the origin and evolution of the lightest elements in the universe – hydrogen and deuterium. In addition, the FUSE satellite will examine the forces and process involved in the evolution of the galaxies, stars and planetary systems by investigating light in the far ultraviolet portion of the electromagnetic spectrum. ["FUSE satellite to study primordial chemical relics of universe evolution," KSC Countdown, June 10, 1999.]

◆ A Boeing Delta 2 rocket lofted four cellular-telephone satellites into orbit Thursday, kicking off a summertime launch surge that calls for five spacecraft-delivery missions in 66 days. The 12-story Delta 2 rocketed away from Cape Canaveral Air Station at 9:49 a.m., propelling the satellite quartet into an orbit 877 miles above Earth. ["Delta kicks off busy summer at Cape," Florida Today, June 11, 1999, p 1A & 2A.]

JUNE 11: About 1,600 workers at Kennedy Space Center went without e-mail Friday to prevent the spread of a destructive computer bug that has infected Windows-based computers in at least 12 countries. NASA's network administrators pulled the plug on the e-mail network mid-morning Friday to prevent the spread of Worm.Explore.Zip, a Trojan Horse computer "worm." The bug is so dangerous that the FBI is urging all users to be cautious when reading their e-mail during the next few days. The FBI's computer crimes unit said it has received reports that the worm infected, damaged or destroyed tens of thousands of computer systems at several major U.S. companies. Boeing Co. first discovered it Thursday, and it had spread to NASA's desktop computers by early Friday. "We deal with viruses pretty frequently, but it's unusual to have a virus as virulent as this one," said Mark Mason, computer security manager at Kennedy Space Center. A dozen to two dozen NASA computers at KSC were infected by the bug, Mason said. NASA's e-mail network was expected to be running by late Friday. KSC's main contractor had to take its entire nationwide e-mail network offline late Thursday to halt the worm's spread. The worm, believed to have originated in Israel, is different from a virus because it spreads by e-mailing itself automatically across the Internet, rather than replicating itself on a computer's hard drive. ["'Worm' infects KSC computers," <u>Florida Today</u>, June 12, 1999, p 1A.]

◆ About 50 protesters gathered Saturday outside Cape Canaveral Air Station to protest the plutonium-powered Cassini spacecraft scheduled to fly by Earth in August. The demonstrators said they were acting in conjunction with protesters all over the world who are involved with the organization Global Network Against Weapons & Nuclear Power in Space. Causing most of the ire was the Cassini spacecraft. Built to study Saturn, its rings and moons, Cassini carries 72 pounds of plutonium to generate electricity for its science instruments. The $3 billion mission was launched from Cape Canaveral in October 1997 on a seven-year journey to Saturn. Gaining speed with each maneuver, the spacecraft is to fly near three planets to reach the ringed giant in 2004. Its path will bring Cassini within 725 miles of Earth on Aug. 17. If the craft were to go off course, NASA's critics worry, it could re-enter the atmosphere accidentally, break apart and release its radioactive plutonium. But officials say the chance of Cassini crashing to Earth is remote – less than one in 1 million. NASA had conducted more than 100 planetary flybys with its spacecraft during the last 40 years. None of them have failed. ["Cassini draws another protest," <u>Florida Today</u>, June 13, 1999, p 1B & 2B.]

JUNE 12: The S0 (S Zero) truss segment, which will become the backbone of the orbiting International Space Station (ISS), arrived at KSC Saturday, June 12, aboard a Super Guppy aircraft. It was transported to the Operations and Checkout Bldg. for processing. The truss segment is a 44- by 15- foot structure weighing 30,800 pounds when fully outfitted and ready for launch. It will be at the center of the ISS 10-truss, girderlike structure that will ultimately extend the length of a footfall field. Eventually the S0 truss will be attached to the U.S. Lab, "Destiny," which is scheduled to be added to the ISS in April 2000. The S0 truss is scheduled to be launched in the Spring of 2001. ["ISS Update," **KSC Countdown**, June 17, 1999.]

JUNE 14: Space Shuttle Status Report, Monday, June 14, 1999. STS-93: Columbia's Helium Signature Test and hypergolic system pressurization are complete. Prelaunch propellant loading is in work today. The orbiter's payload bay has been cleaned and preparations are in work at the pad to support the Chandra payload arrival. Columbia's payload bay doors will be reopened on Wednesday. The Chandra/IUS combination will be placed into the transportation canister on June 18. The payload will be moved to the launch pad on June 19 and installed inside Columbia's cargo bay on June 24. STS-99: Installation of Endeavour's three main engines and modifications to the dome heat shields are complete. Modifications to the final section of line on freon coolant loop No. 2 continue. Interface verification testing on the right-hand orbiter maneuvering system pod is in work. Installation of onboard floodlights is in work as well. STS-103: Removal of Discovery's STS-96 return payload is under way. Checks of the forward reaction control system are ongoing. Later today, workers will begin hypergolic tank valve draining. Routine orbiter receiving inspections continue. STS-101: UHF system checks are complete. Wireless video modifications continue. Mass spectrometer leak checks on the orbiter's fuel cell power plant continue. Bruce Buckingham. (1999). **Kennedy Space Center Space Shuttle Status Report** [Online]. Available E-mail: domo@news.ksc.nasa.gov/subscribe shuttle-status [1999, June 14].]

JUNE 15: Three rocket launch failures that have cost taxpayers at least $3 billion can be traced to human error, the Air Force's top space official told a House subcommittee Tuesday. The panel is investigating malfunctions that include a rocket that exploded after launch Aug. 12, a missile-warning satellite that was launched into the wrong orbit April 9 and a military communications satellite sent into a too-low orbit April 30. All were launched from Cape Canaveral Air Station. The most recent failure seems to have been caused by an error written into the software in the rocket's upper stage, said Keith Hall, assistant secretary of space for the Air Force. "There was the insertion of a decimal point," Hall told the House Technical and Tactical Intelligence subcommittee. "That appears to be the cause." The Air Force is investigating the three cases, all involving Lockheed Martin Titan 4 rockets, but the hearing did not produce details of the human errors that cause the other two failures. Also testifying Tuesday was Pete Aldridge, chief executive officer of The Aerospace Corp., a nonprofit company that provides technical analysis for U.S. and international space programs. The company is responsible for checking Air Force rockets for defects before then blast off. Aldridge said checks and balances are in the system, but somehow the mistake got through. "Quite honestly, the system failed," he said. Failures have not been limited to government launches. Boeing's new Delta 3 heavy-lift launcher has failed twice during the past nine months. The Air Force, Lockheed martin and Boeing are conducting separate investigations of the failures. ["Human error caused three rockets to fail, official says," <u>Florida Today</u>, June 16, 1999, p1A.]

JUNE 16: Payload Status Report, Chandra X-Ray Observatory, Wednesday, June 16, 1999. The Payload Readiness Review for the Chandra X-ray Observatory and the associated Inertial Upper Stage (IUS-27) booster was successfully completed at KSC this afternoon. The payload received a green light to proceed to the next step of launch preparations at the pad. On Friday, June 18, in the Vertical Processing Facility, Chandra will be installed into the payload canister. Then, in the darkness hours of Saturday morning, June 19, it will be transported to Pad 39-B and installed in the payload changeout room. The installation of the Chandra/IUS combination into Columbia's payload bay is currently scheduled to occur on June 24 and will be followed the next day with the electrical connections. On June 28, the interface verification test (IVT) is scheduled to be performed. This will verify the electrical connections and the ability of the astronauts to command the payload from Columbia's flight deck. An end-to-end test will follow on June 30. This will verify the communications path to Chandra, commanding it as if it were in space aboard Columbia. Participating are Chandra's Operations Control Center located in Cambridge, MA, Mission Control at the Johnson Space Center in Houston, the IUS control facility at Sunnyvale, CA and the communications assets of both the Deep Space Network and the Tracking and Data Relay Satellite (TDRS) system. The remaining milestone activity primarily involves the IUS and includes a flight readiness test on July 2, the installation of batteries for flight on July 6-7, and an IUS standalone countdown simulation on July 8. Columbia's payload bay doors are to be closed for flight on July 17 based on the current Chandra/IUS-27 processing schedule which targets a launch on July 20. Bruce Buckingham. (1999). **Payload Status Report, Chandra X-ray Observatory** [Online]. Available E-mail: domo@news.ksc.nasa.gov/subscribe shuttle-status [1999, June 16].]

JUNE 18: Space Shuttle Status Report, Friday, June 18, 1999. STS-93: Workers completed hydraulic close-outs on Columbia's solid rocket boosters on Wednesday. Following yesterday's observance of Super Safety and Health Day, workers today resumed normal launch preparation activities at Launch Pad 39B. In the Vertical Processing Facility, the Chandra/IUS payload is being installed into the payload transportation canister today. However, delivery of the payload to Pad 39B has been rescheduled for Tuesday, June 22. In the payload changeout room at the pad, a roller associated with the telescoping payload ground handling mechanism has failed and must be repaired. Contingency time is available at the pad for Chandra and there is no impact to the targeted July 20 launch date. STS-99: Modifications to the final section of line on freon coolant loop No. 2 continue. Workers are securing Endeavour's recently installed main engines. Preparations to connect the orbital maneuvering system (OMS) cross feed lines are in progress. Next week, Endeavour's external airlock will be installed and OMS nozzle leak checks are scheduled. STS-103: Postflight inspection of Discovery's main engines is complete. Forward reaction control system checks and orbital maneuvering system pod functional tests are under way. Removal of the orbiter's waste recovery system follows next week. STS-101: Installation of Atlantis' left-hand orbital maneuvering system is in work. TACAN antenna installation is also under way. Wireless video modifications continue. Mass spectrometer leak checks on the orbiter's fuel cell power plant continue. Bruce Buckingham. (1999). **Kennedy Space Center Space Shuttle Status Report** [Online]. Available E-mail: domo@news.ksc.nasa.gov/subscribe shuttle-status [1999, June 18].]

JUNE 19: QuikSCAT, a weather forecasting satellite, was launched at 10:15 p.m. EDT on a converted Titan 2 Intercontinental Ballistic Missile from Vandenberg Air Force Base, California. Its science instrument, dubbed SeaWinds, is an antenna that will radiate microwave pulses to measure wind speeds and directions at the surface of the oceans. The data will be put into the National Oceanic and Atmospheric Administration's weather forecasting models. The data also will be studied by scientists looking for long-term trends. ["Titan 2 puts satellite into orbit," <u>Florida Today</u>, June 20, 1999, p 4A.]

◆ NASA officials announced Saturday the launch of $120 million telescope aboard a Boeing Delta 2 rocket this week would be delayed at least a day because work was running behind schedule. Routine prelaunch processing of the Far Ultraviolet Spectroscopic Explorer, or FUSE, at Cape Canaveral Air Station's pad 17A has been slowed because of poor weather, NASA said. Officials will meet Monday to firm up a new launch date, which is tentatively scheduled for Thursday during a window 11:39 a.m. to 12:57 p.m. Once in space, FUSE will seek out chemical remnants of the Big Bang believed to have created the universe 12 to 15 billion years ago. ["Delta 2 launch delayed at least a day, NASA says," <u>Florida Today</u>, June 20, 1999, p 4A.]

JUNE 21: Astronaut Eileen Collins arrived at Kennedy Space Center on Monday to practice for her debut next month as NASA's first woman shuttle commander. The Air Force colonel flew into KSC with the crew of shuttle Columbia, now set for launch July 20 after a year of delays. The mission's goal is to carry a $1.5 billion telescope – called the Chandra X-Ray Observatory – into space to study some of the universe's hottest objects. "There are scientists around the country that (have) been waiting for this mission for years," said Collins, after landing her T-38 jet at NASA's shuttle runway. The crew will be practicing for their launch through Thursday, when they will get strapped into the shuttle and run through the final minutes of countdown in a full dress rehearsal. Originally set for last August, Columbia's mission has been delayed by repairs to the telescope and potential problems with the motor that will propel Chandra to its final orbit after it's released from the spaceship. NASA officials say everything is on schedule for the launch next month. ["Colonel has a date with history," <u>Florida Today</u>, June 22, 1999, p 1A.]

JUNE 22: Delta/FUSE Launch Weather Forecast for June 24, 1999. Forecast issued L-2 days, June 22, 1999. Forecast: Partly cloudy with developing afternoon showers and thunderstorms after the onset of the seabreeze. The weather concerns grow as the launch window enters afternoon hours. During the launch window (11:39 a.m. -12:57 p.m.) the forecast is: Clouds: 3,000 feet scattered, 3/8 sky coverage; 15,000 feet broken, 5/8 sky coverage; Visibility: 10+ miles; Wind: SW/5-10 knots becoming SE/10-15 knots; Temperature: 84 degrees; Relative Humidity: 80%; Probability of launch weather criteria violation: 40%; Probability with 24 hour delay: 40%; Probability with 48 hour delay: 40%; Forecast prepared by USAF 45th Weather Squadron. Bruce Buckingham. (1999). **Delta/FUSE Launch Weather Forecast for June 24, 1999** [Online]. Available E-mail: domo@news.ksc.nasa.gov/subscribe shuttle-status [1999, June 22].]

◆ In 50 years, "space cruises" from Kennedy Space Center could compete against ocean cruise lines for tourism dollars, KSC Director Roy Bridges predicted Tuesday. "Instead of coming down here to go on a cruise on the Disney Magic, we're going to be coming down here to get on a space liner to go for a cruise in low-Earth orbit," Bridges said. Also, such courier companies as Federal Express might leave KSC to deliver commercial cargo to other continents such as Australia. Travel time? "A little over an hour," Bridges said. Bridges' comments were made to about 200 area business leaders at the center's annual Community Leaders breakfast. The meeting is a briefing for area economic-development leaders on upcoming developments at KSC. As a whole, KSC helps generate close to $1 billion in direct economic benefit to the state. And its Visitor Complex is Florida's fifth-ranking tourist destination. Bridges' predictions on commercial space flight from KSC are based on new technologies that will someday offer low-Earth orbit travel resembling landings and takeoffs from a standard airport. The passenger vessel would be the space equivalent of a Boeing 777 jumbo aircraft. No word from him, though, on the cost of a space cruise. "The good thing about this for KSC is that this spaceport here in Florida ought to be the hub of all planetary departures," Bridges said. Chuck Rowland, executive director of the Canaveral Port Authority, said there's no reason to fear an ocean cruise vs. space cruise matchup in Brevard County. Cruise lines already tie visits to KSC into some of their vacation packages. ["Space cruises from KSC predicted," <u>Florida Today</u>, June 23, 1999, p 12C.]

JUNE 23: Space Shuttle Status Report, Wednesday, June 23, 1999. STS-93: Yesterday, workers disabled the drive train for truck No. 1 on the Rotating Service Structure (RSS) at Launch Pad 39B to accommodate RSS rotation tests this morning. With those tests successfully completed, RSS operations will proceed normally using only truck No. 2 . If all continues to go well, the Chandra/IUS payload will arrive at Launch Pad 39B early Thursday morning for installation into the payload changeout room on the RSS. The payload will then be installed into Columbia's payload bay no earlier than Monday. Last Monday night one of the two RSS trucks that extend and retract the RSS away from the Shuttle failed to operate properly. Because the RSS did not fully retract and stopped after moving only four inches, the Chandra/IUS payload was not transferred to the launch pad. The RSS must be retracted for a payload to be installed into the payload changeout room. A hot fire test of auxiliary power unit No. 2 has been successfully completed. Terminal Countdown Demonstration Test activities continue, but Thursday's dress rehearsal will now start two hours late culminating with a simulated main engine cut-off at about 1 p.m. Columbia remains targeted for launch no earlier than July 20. STS-99: Leak checks on Endeavour's recently modified freon coolant loop No. 2 are in progress. Workers are installing the orbiter external airlock and the payload premate test is in work. Later this week, technicians will connect the orbiter's auxiliary power unit drain lines. STS-103: Removal of Discovery's main engines is complete. Preparations for orbiter maneuvering system functional tests are under way. This week workers will remove the orbiter's forward reaction control system and replace fuel cell No. 2. Removal of the transfer tunnel adapter is also scheduled this week. STS-101: Atlantis' left-hand orbital maneuvering system checks are in work. TACAN antenna installation, wireless video modifications, and mass spectrometer leak checks on the orbiter's fuel cell power plant continue. Bruce Buckingham. (1999). **Kennedy Space Center Space Shuttle Status Report** [Online]. Available E-mail: domo@news.ksc.nasa.gov/subscribe shuttle-status [1999, June 23].]

JUNE 24: Space Shuttle Status Report, Thursday, June 24, 1999. STS-93: The Chandra/IUS payload arrived at Launch Pad 39B today at about 3 a.m. and transfer into the Payload Changeout Room began at about 6 a.m. Managers are now targeting Sunday for payload installation into Shuttle Columbia's payload bay. Later today, workers begin calibrating the orbiter's inertial measurement units and will conduct leak checks on the orbiter midbody umbilical unit. The Terminal Countdown Demonstration Test concluded today with the five-member flight crew participating in a launch day dress rehearsal. After a simulated main engine cutoff at 1 p.m., crew members practiced emergency egress procedures from Columbia's crew module. The astronauts will return to Houston, TX later today and resume their final preflight training. STS-99: Endeavour's external airlock was installed yesterday. Leak checks on freon coolant loop No. 2 continue. The payload premate test is ongoing and later today technicians will connect the orbiter's auxiliary power unit drain lines. STS-103: Discovery's orbital maneuvering system is undergoing functional tests. Functional checks of the orbiter's forward reaction control system are also under way and the FRCS is slated for removal early next week. Workers are preparing to remove all three of Discovery's fuel cells. STS-101: Installation of Atlantis' TACAN antennas is complete. Checks on the left-hand orbital maneuvering system pod continue. Wireless video modifications and mass spectrometer leak checks on the orbiter's fuel cell power plant continue. Managers plan to install the orbiter's external airlock early next week. Bruce Buckingham. (1999). **Kennedy Space Center Space Shuttle Status Report** [Online]. Available E-mail: domo@news.ksc.nasa.gov/subscribe shuttle-status [1999, June 24].]

◆ Payload Status Report, Chandra X-Ray Observatory, Thursday, June 24, 1999. The Chandra X-ray Observatory with the attached Inertial Upper Stage booster arrived at launch pad 39-B this morning at 3:50 a.m. It left the Vertical Processing Facility last night at 12:30 a.m. inside the payload canister riding atop the payload transporter. It is now being removed from the payload canister and installed inside the payload changeout room at the pad. The installation of the Chandra/IUS combination into Columbia's payload bay is currently scheduled to occur on Sunday, June 27 and will be followed the next day with the electrical connections. On June 30, the interface verification test (IVT) is scheduled to be performed. This will verify the electrical connections and the ability of the astronauts to command the payload from Columbia's flight deck. An end-to-end test will follow on July 2. This will verify the communications path to Chandra, commanding it as if it were in space aboard Columbia. Participating are Chandra's Operations Control Center located in Cambridge, MA, Mission Control at the Johnson Space Center in Houston, the IUS control facility at Sunnyvale, CA and the communications assets of both the Deep Space Network and the Tracking and Data Relay Satellite (TDRS) system. The remaining milestone activity primarily involves the IUS and includes a flight readiness test on July 5, the installation of batteries for flight on July 7-8, and an IUS standalone countdown simulation on July 9. Columbia's payload bay doors are to be closed for flight on July 17 based on the current Chandra/IUS-27 processing schedule which targets a launch on July 20. Bruce Buckingham. (1999). **Payload Status Report, Chandra X-Ray Observatory** [Online]. Available E-mail: domo@news.ksc.nasa.gov/subscribe shuttle-status [1999, June 24].]

◆ NASA's newest space telescope is flying 477 miles above Earth today, giving astronomers a new tool with which to study the explosive and mysterious birth of the universe. It was launched on the first leg if its mission atop a 12-story Boeing Delta 2 rocket that blasted off Thursday from Cape Canaveral Air Station. The telescope is expected to help astronomers understand the life cycles of galaxies and stars. "We're off to a great start," said NASA mission manager David Mengers. "The spacecraft is now on its way on a three-year study to shed light on the origins of the universe," added NASA launch commentator Lisa Malone. A cosmic time machine of sorts, NASA's Far Ultraviolet Spectroscopic Explorer, or FUSE, will seek the chemical remnants of the Big Bang, the primordial explosion scientists think created the universe 12 billion to 15 billion years ago. The FUSE mission is the latest in a series being staged as part of NASA's Origins Program. Its aim is to study the birth and evolution of stars and galaxies and to search for planets that might harbor life. ["Successful launch has FUSE up, running," <u>Florida Today</u>, June 25, 1999, p 1A.]

JUNE 28: Space Shuttle Status Report, Monday, June 28, 1999. STS-93: Shuttle Columbia's prelaunch processing continues on schedule for a target launch date of July 20. Last Friday, Columbia's payload bay doors were opened in preparation for payload installation. Sunday, the Chandra/IUS payload was transferred into Columbia's payload bay, and final connections are in work today. The orbiter/payload Interface Verification Test follows on Wednesday. Later this week, workers will verify the Shuttle cavity purge lines and conduct leak checks on the mated orbiter midbody umbilical unit. STS-99: Endeavour's auxiliary power unit fuel and drain line connections are complete. Leak checks on freon coolant loop No. 2 concluded last week and servicing begins Wednesday. The orbiter's main engine heatshields are being installed today, and the payload premate test is ongoing. STS-

103: Functional tests on Discovery's forward reaction control system (FRCS) are complete, and the FRCS will be removed tonight. Orbital maneuvering system testing continues. Technicians are servicing the upper hatch on Discovery's docking system today. Preparations are under way to remove and replace all three Shuttle fuel cells, beginning with fuel cell No. 2 this week. STS-101: Checks on Atlantis' left-hand orbital maneuvering system pod continue. Wireless video modifications and mass spectrometer leak checks on the orbiter's fuel cell power plant continue. Workers will also install the orbiter's airlock this week. Bruce Buckingham. (1999). **Kennedy Space Center Space Shuttle Status Report** [Online]. Available E-mail: domo@news.ksc.nasa.gov/subscribe shuttle-status [1999, June 28].]

◆ Florida has received high marks in its effort to be the launch site for the next generation spaceship while facing competition from 14 other states. A three-person team representing Lockheed Martin, designer of the proposed VentureStar vehicle, met with state space and economic leaders in Orlando on Monday to offer feedback on Florida's launch and operations proposal submitted last September. Lockheed Martin will choose a pair of U.S. launch sites for the coveted $5 billion project in early 2001. "It was a good meeting," said Anthony Jacob, a VentureStar business development executive at Lockheed Martin. "All of these people are very smart about the launch business." Ed O'Connor, director of Spaceport Florida Authority, the state agency responsible for attracting space industry, said Florida officials were optimistic but they weren't going to take anything for granted. The states are being graded in four major areas: safety of flights for surrounding populations; how well the launch site serves commercial markets by offering access to different types of orbits; costs of operations; cost of building the launch site and support facilities. Lockheed Martin and Florida officials wouldn't discuss the specifics of Monday's briefing, but a participant told *The Orlando Sentinel* that the state's bid scored above average in most key categories. Florida rated first in two of the most crucial categories, safety risk and the project tax burden of operating in the state, *The Sentinel* reported. Florida's proposal designates an area at Kennedy Space Center, north of the space shuttle's Launch Complex 39, as the VentureStar launch site with the shuttle runway serving as the landing strip. If VentureStar becomes a reality, the single-stage reusable launcher would revolutionize the space industry by dramatically lowering the cost of putting cargo in orbit. It blasts off without shedding stages or boosters, then lands like an airplane. Test flights of a $1.2 billion prototype are scheduled to begin in July 2000. The first VentureStar flights would begin no earlier than 2005. ["State gains edge in luring VentureStar," Florida Today, June 30, 1999, p 6B.]

DURING JUNE: NASA's Lunar Prospector will end its 18-month mission on July 31 when ground controllers attempt a direct hit of a permanently shadowed crater near the Moon's south pole. ["World news roundup," Aviation Week & Space Technology, June 7, 1999, p 20.]

◆ Ralph Roe, Jr., NASA Space Shuttle launch director at Kennedy Space Center, is moving to Houston, TX and into a new position as manager of the Space Shuttle Vehicle Engineering Office at Johnson Space Center. In this position, Roe will direct the design, production, and testing of Space Shuttle orbiters, associated government-furnished equipment, the remote manipulator system, software, avionics, and flight crew equipment. Roe joined NASA at KSC in 1983 and has served in several senior technical and managerial positions. ["KSC shuttle launch director accepts new position at JSC," **The Brevard Technical Journal**, June 1999, p 4.]

JULY

JULY 1: Space Shuttle Status Report, Thursday, July 1, 1999. STS-93: At Launch Pad 39B, workers continue to prepare Shuttle Columbia and the Chandra/Inertial Upper Stage (IUS) payload for a target launch date of July 20. Over the weekend, Shuttle and payload workers will support planned payload activities. Next Tuesday, workers will install mass memory unit (MMU) No. 1. Loading of MMU No. 1 and No. 2 commences the following Thursday. The payload Interface Verification Test that validates the payload's electrical connections to the orbiter continues on schedule. Today, workers begin IUS flight battery installation, and on Friday, the payload end-to-end test begins. STS-99: Installation of Endeavour's three main engines is complete. Payload premate testing is also complete and servicing of freon coolant loop No. 2 concluded yesterday. Installation of the Shuttle main engine heatshields is in work. Later this week, technicians will begin ammonia system servicing. STS-103: Discovery's orbital maneuvering system continues to undergo functional tests. Removal of all three Shuttle fuel cells proceeds on schedule with removal of fuel cell No. 2 already complete. STS-101: Installation of Atlantis' forward reaction control system is in progress. Wireless video modifications and mass spectrometer leak checks on the orbiter's fuel cell power plant continue. Installation of the orbiter's external airlock is slated to begin next week. Bruce Buckingham. (1999). **Kennedy Space Center Space Shuttle Status Report** [Online]. Available E-mail: domo@news.ksc.nasa.gov/subscribe shuttle-status [1999, July 1].]

◆ The underwater salvage expert who found Gus Grissom's Mercury capsule at the bottom of the Atlantic two months ago headed back out to sea Thursday to lift the 38-year-old spacecraft from its watery grave. "It will be a big relief" once the capsule is aboard ship, Curt Newport said. "I'll feel like the weight of a Mercury capsule's been taken off my shoulders." Newport and his team discovered Liberty Bell 7 on May 1 in 3-mile-deep water about 300 miles southeast of Cape Canaveral. But they were forced to leave it there when the cable to their robotic recovery vessel snapped in rough seas. It took five weeks for Oceaneering International, Inc. to build another recovery vessel. The Houston company also supplied a bigger ship for this trip, which is expected to last 1 ½ to two weeks. Newport plans to retrieve the recovery vessel first, then go after Liberty Bell 7, the only U.S. manned spacecraft lost after a successful mission. ["Expedition under way to get Mercury capsule," <u>Florida Today</u>, July 2, 1999, p 7A.]

JULY 2: Payload Status Report, Chandra X-Ray Observatory, Friday, July 2, 1999. Among STS-93 payload activities this past week, the installation of the Chandra/IUS combination into Columbia's payload bay was performed as scheduled on Sunday, June 27. This was followed with the start of the interface verification test (IVT) on Wednesday, June 30. The test was successfully completed last night with no significant issues or concerns. The IVT verifies the electrical connections with the orbiter and the ability of the astronauts to command the payload from Columbia's flight deck. Also on Thursday, the IUS computer memory battery was installed. The end-to-end test began this morning. This is an 18-hour test to check the communications path to Chandra, commanding it as if it were in space aboard Columbia. Participating are Chandra's Operations Control Center located in Cambridge, MA, Mission Control at the Johnson Space Center in Houston, the IUS control facility at Sunnyvale, CA, and the communications assets of both the Deep Space Network and the Tracking and Data Relay Satellite (TDRS) system. Over this weekend, the Chandra telescope's flight batteries are scheduled to be charged. The remaining milestone activity primarily involves the IUS and includes a flight readiness test on July 5, the installation of flight batteries on July 7-8, and an IUS stand-alone countdown simulation on July 9. Columbia's payload bay doors are to be closed for flight on July 17 based on the current Chandra/IUS-27 processing schedule which targets a launch on July 20. Bruce Buckingham. (1999). **KSC Chandra X-Ray Telescope Status Report** [Online]. Available E-mail: domo@news.ksc.nasa.gov/subscribe shuttle-status [1999, July 2].]

JULY 6: Construction of NASA's troubled International Space Station might have hit another snag after a Russian rocket failure, agency officials said Tuesday. The accident involved an unmanned Proton rocket, which malfunctioned Monday and crashed in central Asia while trying to launch a Russian military communications satellite. The mishap raises concerns about the scheduled Nov. 12 launch of a Proton carrying Russia's Zvezda service module for the International Space Station. The module, which is more than a year behind schedule, will serve as the initial living quarters for the outpost. Delays in getting the piece into orbit have effectively put the $60 billion project on hold. The most recent investigation into a Proton mission failure in 1997 took about four months to complete. "Until the investigation is completed, we really won't know what impact, if any, this will have on the Service Module's launch," NASA spokeswoman Debra Rahn said. Zvezda's launch would be the fourth of 88 American and Russian spaceflights needed to build, outfit and staff the orbiting laboratory by 2005. The construction plan would be halted until Zvezda arrives at the infant station because the module must be in place before future pieces of the station can be attached. ["Russian rocket crashes in Asia," Florida Today, July 7, 1999, p 1A.]

JULY 8: A high-powered X-ray telescope and NASA's first woman shuttle commander are set for launch July 20. NASA managers Thursday officially cleared the flight of shuttle Columbia, which will be led by Air Force Col. Eileen Collins. The second shuttle flight of the year, Columbia's mission is to carry the Chandra X-Ray Observatory into space. ["NASA clears Columbia for July 20 launch," Florida Today, July 9, 1999, p 1A.]

◆ Apollo 12 astronaut Pete Conrad, the 3rd man to walk on the moon, died of injuries suffered in a motorcycle accident on Thursday in Southern California. He was 69. Flags flew at half-staff at Kennedy Space Center and other NASA installations Friday in observance of his death. ["Astronaut Pete Conrad, 69, dies after motorcycle crash," The Orlando Sentinel, July 10, 1999, p A-1.]

JULY 10: A Delta II rocket carrying 4 cellular-telephone satellites lifted off from Cape Canaveral Air Station at 4:45 a.m. Saturday. After a few days of checkouts, ground controllers will move the satellites into orbits 878 miles above Earth. The satellites will eventually be part of a global wireless-communications network. ["A giant leap for cell phones," The Orlando Sentinel, July 11, 1999, p A-3.]

JULY 11: As the memory of the 1986 space shuttle Challenger disaster fades and competing specialty license plates spring up, the Astronaut Memorial Foundation has fallen on lean times. The $15.7 million organization lost $131,026 in fiscal year 1997-98, according to the most recent internal audit AMF files with the state. Losses are expected to continue through this fiscal year. AMF president Stephen Feldman insists the losses, due in large part to declining tag sales, don't threaten the memorial to dead astronauts at Kennedy Space Center or AMF's other mission, teaching technology to educators through its Center for Space Education. AMF likely will post another year of losses in its next financial report, but it expects to have $115,000 in cash reserves by June 2000. AMF is banking that a new Challenger specialty tag it unveiled last month will help significantly. ["Astronaut foundation funds falter," Florida Today, July 12, 1999, p 1A & 2A.]

JULY 12: Space Shuttle Status Report, Monday, July 12, 1999. STS-93: Following last Thursday's Flight Readiness Review, Shuttle managers announced July 20 as the official launch date for STS-93. Truck No. 1 on the Rotating Service Structure at Launch Pad 39B, has been repaired and is ready to support standard prelaunch activities. Last Friday, engineers completed a simulated countdown test for the payload's inertial upper stage and workers conducted routine voltage tests as well. Columbia's aft compartment closeouts are in work this week. Flight crew equipment early stow occurs tomorrow along with preparations for ordnance installation. STS-99: Integrated orbiter hydraulic testing concludes later this week. Endeavour's midbody close-outs are ongoing. Orbiter fluid servicing is also under way. Preparations are under way to receive the SRTM payload into the OPF and then install it into the orbiter's payload bay next Tuesday. The orbiter/payload interface verification test is slated for July 27. STS-103: Replacement of Discovery's three fuel cells is complete. Over the weekend, workers drained residual propellant from the orbiter's right-hand orbital maneuvering system engine. Workers are scheduled to install a new right-hand orbital maneuvering engine this week. Removal of Discovery's robot arm from the payload bay begins tomorrow. Main propulsion system leak checks continue and ammonia system servicing begins this week. STS-101: Last Friday, workers closed Atlantis' payload bay doors to accommodate optical payload bay fit-checks over the weekend. The payload bay doors will be opened again today. Managers plan to install the external airlock tomorrow. Next week, workers will begin preparing Atlantis for a brief storage period in the Vehicle Assembly Building. Atlantis is scheduled to roll over to VAB high bay 2 on July 23 and remain there until it transfers to OPF bay 2 on Aug. 16. Atlantis is leaving OPF bay 3 to make room for orbiter Columbia's return from STS-93 and moves into OPF bay 2 following Endeavour's departure for the VAB. Bruce Buckingham. (1999). **Kennedy Space Center Space Shuttle Status Report** [Online]. Available E-mail: domo@news.ksc.nasa.gov/subscribe shuttle-status [1999, July 12].]

◆ A tiny but efficient orbiter that will map Mercury and a spacecraft that will blast a hole into a comet so scientists can see what is inside are two latest missions on NASA's new cheaper-but-better Discovery Program agenda. The Mercury Space Environment Geochemistry and Ranging mission (MESSENGER) will carry seven miniaturized instruments to look for water, a magnetic field and other interesting phenomena on the planet, and will be the first spacecraft to visit Mercury in more than 30 years. The spacecraft, scheduled to be launched in 2004, will also send back the first global images of mercury, the closest planet to the sun. The Deep Impact mission will shoot a 1,100-pound copper projectile into comet P/Temple 1, blasting out a crater as big as a football field and as deep as a seven-story building. It will carry a camera and infrared spectrometer to study the resulting icy debris and peer inside the comet. ["Craft will probe secrets of Mercury, comet on cheap," The Orlando Sentinel, July 13, 1999, p A-3.]

JULY 13: Payload Status Report, Chandra X-Ray Observatory, Friday, July 13, 1999. Payload prelaunch preparations continue on schedule at Pad 39-B to prepare for the liftoff of Columbia on STS-93. The Chandra X-ray Observatory and the attached Inertial Upper Stage (IUS) were each declared ready for launch at the completion of the flight readiness review held at KSC last Thursday. Meanwhile, the Chandra operations team successfully completed their final mission dress rehearsal this weekend and is ready to support the post-launch activities. Participating in the flight simulation were Chandra's Operations Control Center in Cambridge, MA, Mission Control at Johnson Space Center in Houston, TX, the Inertial Upper Stage control facility in Sunnyvale, CA, and the communications assets of both the Deep Space Network and the Tracking and Data Relay Satellite System. The separate IUS stand-alone countdown simulation was successfully completed last Friday, July 9. This put the IUS through all countdown activities as if it were launch day. Connection flight ordnance associated with IUS separation events and for Chandra's fuel system is scheduled for Wednesday, July 14. This will be followed with payload closeouts on Thursday which continue through Saturday. This includes removal of red tag items which are items not to be flown on the payload, the disconnection of purge lines, contamination inspections, configuration checks and the taking of closeout photos. Columbia's payload bay doors will be closed for flight on Saturday, July 17. The IUS will be powered-on July 19 during the countdown at 3:16 a.m. (L-21 hours, 20 minutes. Bruce Buckingham. (1999). **KSC Chandra X-Ray Telescope Status Report** [Online]. Available E-mail: domo@news.ksc.nasa.gov/subscribe shuttle-status [1999, July 13].]

◆ Space Shuttle Status Report, Tuesday, July 13, 1999. STS-93: At Launch Pad 39B, preparations continue to go well for Columbia's July 20 launch date. Yesterday, workers replaced an audio communication component onboard Columbia and retesting is in work today. Close-outs of the orbiter aft compartment are ongoing and Shuttle ordnance installation begins tonight. STS-99: Integrated orbiter hydraulic testing concludes later this week. Endeavour's payload bay closeouts are in progress. Auxiliary power unit controller checks are in work and potable water servicing is under way. Preparations are under way to receive the SRTM payload into the OPF and then install it into the orbiter's payload bay next Tuesday. The orbiter/payload interface verification test begins next week. STS-103: Workers are scheduled to install a new right-hand orbital maneuvering engine this week. Removal of Discovery's robot arm from the payload bay begins today. Main propulsion system leak checks continue and ammonia system servicing begins this week. STS-101: Atlantis' payload bay doors were opened today. Managers now plan to install the orbiter's external airlock on Friday. Next week, workers will begin preparing Atlantis for a brief storage period in the Vehicle Assembly Building. Atlantis is scheduled to roll over to VAB high bay 2 on July 23 and remain there until it transfers to OPF bay 2 on Aug. 16. Atlantis is leaving OPF bay 3 to make room for orbiter Columbia's return from STS-93 and moves into OPF bay 2 following Endeavour's departure for the VAB. Bruce Buckingham. (1999). **Kennedy Space Center Space Shuttle Status Report** [Online]. Available E-mail: domo@news.ksc.nasa.gov/subscribe shuttle-status [1999, July 13].]

◆ The 45th Space Wing will get a new commander next month. Brig. Gen. Donald P. Pettit will take command of Patrick Air Force Base's main unit in late August. He replaces Brig. Gen. Randall Starbuck, the wing's commander since March 1997. ["Patrick's Space Wing to get new leader in August," Florida Today, July 14, 1999, p 1A.]

◆ The much-delayed launch of the newest U.S. weather satellite is being postponed again, this time because of the orbital dance between the Earth and sun, officials said Tuesday. The $250 million GOES-L spacecraft was scheduled for launch in May. However, its ride on a Lockheed Martin Atlas 2A rocket has been grounded this summer because of investigations into a series of rocket failures. The launch has been delayed until at least October at the request of the satellite's ground controllers. ["Weather satellite's launch delayed," Florida Today, July 14, 1999, p 1A.]

◆ A underwater salvage team recovered the dye canister from Gus Grissom's Mercury capsule Tuesday and hopes to finish raising the capsule itself today. ["Capsule retrieval," Florida Today, July 14, 1999, p 1A.]

◆ Ticket requests for a Friday banquet honoring the first men on the moon have gone through the roof, forcing NASA officials to turn away about 200 people. The Kennedy Space Center gala, which celebrates the 30th anniversary of the Apollo 11 journey to the moon, will seat 950 people, mostly former workers in the space program. But more than 1,100 checks for $45 were mailed in, and refunds were sent out to the last 200 that came in, said Hugh Harris, a former KSC spokesman who is one of the event's organizers. The KSC fire marshal said allowing more than 950 people would be a safety hazard, considering all the dinner tables in the room, Harris said. Harris was not surprised by the banquet's popularity. "This is the last year of the century, and this is the single biggest achievement of the century," he said. "I'm sure (those turned away) are very disappointed, but I'm sure they understand." Astronauts expected to attend include Neil Armstrong and Buzz Aldrin, the Apollo 11 astronauts who were the first men on the moon, and Apollo astronauts Wally Schirra, Walt Cunningham, Charles Duke and Eugene Cernan. The celebration will last two days. ["Apollo 11's 30th anniversary banquet sold out," **Florida Today**, July 15, 1999, p 1B.]

JULY 15: Space Shuttle Status Report, Thursday, July 15, 1999. STS-93: At Launch Pad 39B, workers completed ordnance installation this morning and close-outs are in work. Retests of the recently replaced audio central control unit in Columbia's crew compartment were successfully completed yesterday. Orbiter aft compartment and airlock close-outs continue. Preparations are under way in the Launch Control Center to start the countdown clock at 10 p.m. tomorrow. The STS-93 flight crew arrives at KSC tomorrow at about 7 a.m. STS-99: Technicians have completed leak checks on Endeavour's external airlock interface line and airlock checks follow today. Payload bay close-outs continue and gaseous nitrogen servicing is ongoing. Auxiliary power unit controller checks continue and potable water servicing proceeds on schedule. Preparations are under way to receive the SRTM payload into the OPF and then install it into the orbiter's payload bay next Tuesday. The orbiter/payload interface verification test begins next week. STS-103: Servicing of Discovery's waste management system is complete. Retests of the orbiter's three fuel cells are complete and good. Main propulsion system leak checks continue and orbital maneuvering system pod functional testing is in progress. Today, workers are removing Discovery's robot arm. This week technicians will install the orbiter's main landing gear. The right-hand orbital maneuvering engine is still being drained of residual fuel in preparation for next week's planned replacement. STS-101: Preparation is under way to install Atlantis' external airlock tomorrow. Wireless video system modifications continue. Technicians are installing orbiter window No. 1 this week and water spray boiler checks are slated this week. Next week, workers will begin preparing Atlantis for a brief storage period in the Vehicle Assembly Building. Atlantis is scheduled to roll over to VAB high bay 2 on July 23 and remain there until it transfers to OPF bay 2 on Aug. 16. Atlantis is leaving OPF bay 3 to make room for orbiter Columbia's return from STS-93. Bruce Buckingham. (1999). **Kennedy Space Center Space Shuttle Status Report** [Online]. Available E-mail: domo@news.ksc.nasa.gov/subscribe shuttle-status [1999, July 15].]

◆ NASA plans to make a little history on the 30th anniversary of the Apollo 11 moon landing, launching the nation's first woman commander of a space mission. Shuttle Columbia skipper Eileen Collins and her crewmates say Neil Armstrong's boot prints on the moon still serve as an inspiration for those following in his footsteps. "It's really kind of a neat feeling to know that we're going to be a part of that history and continuing on with the space program the way we are," Collins said. Collins and her crew are scheduled to blast off at 12:36 a.m. Tuesday from Kennedy Space Center's launch pad 39B, 1½ miles from where Apollo 11 embarked on its historic mission. Columbia's mission is to deploy a $1.5 billion X-ray astronomy observatory from the cargo bay of the shuttle, giving astronomers a powerful new tool to study exploding starts, quasars and black holes. ["Columbia to lift off 30 years after first landing on moon," <u>Florida Today</u>, July 16, 1999, p 8A.]

JULY 16: Space Shuttle Status Report, Friday, July 16, 1999. STS-93: Preparation for the launch of Shuttle Columbia on July 20 continues on schedule at Launch Pad 39B. Workers have completed orbiter aft compartment close-outs and are preparing to pickup the launch countdown tonight at 10 p.m. Payload close-outs are in work in preparation for tomorrow's payload bay door closure. Flight crew equipment stowage is on going. The five-member STS-93 flight crew arrived at KSC today at about 7 a.m. STS-99: Technicians have completed functional tests on Endeavour's landing gear. Payload bay close-outs continue and potable water servicing is ongoing. Leak checks of the crew cabin and external airlock are scheduled today. Preparations are under way to receive the SRTM payload into the OPF and then install it into the orbiter's payload bay next Tuesday. The orbiter/payload interface verification test begins next week. STS-103: Removal of Discovery's robot arm is complete. Main propulsion system leak checks continue and draining of the right-hand orbital maneuvering system (OMS) engine is in progress. Later this week, workers will begin OMS engine disconnects. Discovery's main landing gear is being installed today. Drag chute installation begins Saturday. STS-101: Installation of Atlantis' external airlock is under way. Wireless video system modifications resume tomorrow. Technicians are installing orbiter window No. 1 this week and water spray boiler checks are scheduled this week. Freon coolant loop functional testing is in progress. Next week, workers will begin preparing Atlantis for a brief storage period in the Vehicle Assembly Building. Atlantis is scheduled to roll over to VAB high bay 2 on July 23 and remain there until it transfers to OPF bay 2 on Aug. 16. Atlantis is leaving OPF bay 3 to make room for orbiter Columbia's return from STS-93. Bruce Buckingham. (1999). **Kennedy Space Center Space Shuttle Status Report** [Online]. Available E-mail: domo@news.ksc.nasa.gov/subscribe shuttle-status [1999, July 16].]

◆ KSC and the 45th Space Wing at Patrick AFB are being given the Hammer Award for establishing J-BOSC in 1998. The award is a special recognition by Vice President Al Gore of teams of federal employees who have made significant contributions supporting the principles of the National Partnership for Reinventing Government. The award will be presented by Morley Winograd, director of the National Partnership. The award ceremony is being held July 16 at 4 p.m. at the IMAX2 Theater. NASA Administrator Daniel Goldin and Commander of the Air Force Space Command Gen. Richard B. Myers are expected to speak. ["KSC, 45th Space Wing get Hammer Award for J-BOSC accomplishments," **KSC Countdown**, July 15, 1999.]

◆ When Neil Armstrong, Buzz Aldrin and Michael Collins set sail 30 years ago today aboard Apollo 11, the moon was still remote and mysterious. Armstrong's first step on the lunar surface four days later signaled not only a giant leap for mankind but U.S. space supremacy. Astronauts, NASA workers and other luminaries spent most of the day Friday reliving man's first lunar-landing mission, which blasted off from Kennedy Space Center on July 16, 1969. The day closed with a gala banquet for 950 people at the Apollo/Saturn 5 Center – a KSC museum dedicated to the moon rocket. Michael Collins, who flew with Armstrong and Aldrin to the moon, declined an invitation to attend. The former astronauts walked into a tourist exhibit of NASA's old firing room consoles for a new conference. The 68-year-old Armstrong said his "gut feeling" when he lifted off with Aldrin and Collins is that they had a 90 percent chance of returning safely to Earth and that he and Aldrin had a 50 percent chance of landing successfully on the moon. ["Is love affair with the moon over?" **The Orlando Sentinel**, July 16, 1999, p A-1 & A-4. "Apollo astronauts relive history," **Florida Today**, July 17, 1999, p 1A & 2A.]

JULY 17: Space Shuttle Status Report, Saturday, July 17, 1999. STS-93: At 10 p.m. yesterday the countdown for launch of Space Shuttle Columbia began on schedule at the T-43 hour mark. Early stowage of the flight crew equipment into the orbiter's crew compartment concluded yesterday. Last night the STS-93 astronauts conducted a final inspection of the Chandra X-ray Observatory and this morning workers closed Columbia's payload bay doors for flight. At about 6 p.m. today, the 8-hour operation begins to load cryogenic reactants into the orbiter's onboard fuel cell storage tanks. Weather forecasters indicate a 30 percent probability that weather could prohibit Tuesday's launch attempt. The primary concern is the chance of coastal showers. The forecast calls for clouds to be scattered at 3,000 and 25,000 feet; visibility at 7 miles; wind out of the east at 8 knots peaking to 12 knots; temperature at 79 degrees F and a chance of coastal showers. Crew for mission STS-93: Commander (CDR): Eileen Collins; Pilot (PLT): Jeff Ashby; Mission Specialist (MS1): Catherine Coleman; Mission Specialist (MS2): Steven Hawley; Mission Specialist (MS3): Michel Tognini; Summary of launch day crew activities: Monday, July 19 - 7 p.m. Crew wake up; *7:30 p.m. 8:11 p.m. Crew breakfast/photo CDR, PLT, MS2 weather briefing; 8:11 p.m. 8:21 p.m. *8:30 p.m. *8:51p.m. *9:21 p.m. *10:36 p.m. MS1and MS3 don launch and entry suits CDR, PLT, MS2 don launch and entry suits Crew suit up photo Depart for Launch Pad 39B Arrive at white room and begin ingress Close crew hatch; Tuesday, July 20 *12:36 a.m. Launch. Bruce Buckingham. (1999). **Kennedy Space Center Space Shuttle Status Report** [Online]. Available E-mail: domo@news.ksc.nasa.gov/subscribe shuttle-status [1999, July 17].]

JULY 18: Space Shuttle Status Report, July 18, 1999. STS-93: Operations continue on schedule for the launch of Shuttle Columbia at 12:36 a.m. Tuesday, July 20. Columbia's payload bay doors have been closed for flight and no additional pre-launch work on the Chandra X-ray Observatory is scheduled. Loading of the on-board cryogenics was concluded earlier this morning, and off-loading of the extra cryogenics not needed for use in the orbiter's fuel cells during the mission will continue until early afternoon today. Following this, the orbiter mid-body umbilical unit will be retracted and final pre-launch preparations of the three Shuttle main engines will commence. Activation of the orbiter's communications system will begin at about 11 p.m. today. Also, work to complete final stowage of the flight crew's equipment will begin soon after midnight tonight, as will preparations for the retraction of the rotating service structure at 4 a.m. tomorrow. The three-hour operation to load the external tank with 500,000 gallons of liquid oxygen and liquid hydrogen will begin at about 3:46 p.m. Monday. Following this operation, the crew will be awakened at about 7 p.m., they will be seated for their pre-launch meal at 7:30 p.m., and depart for Launch Pad 39B at 8:51 p.m. Weather forecasters continue to indicate a 30 percent probability that weather could prohibit Tuesday's launch attempt. The single concern is for a chance of coastal showers. The forecast calls for clouds to be scattered at 3,000 and 25,000 feet; visibility at 7 miles; winds out of the east at 8 knots peaking to 12 knots; temperature at 80 degrees F and humidity 86 percent. Bruce Buckingham. (1999). **Kennedy Space Center Space Shuttle Status Report** [Online]. Available E-mail: domo@news.ksc.nasa.gov/subscribe shuttle-status [1999, July 18].]

◆ After two days of searching and holding out hope of a miraculous survival, authorities conceded late Sunday that John F. Kennedy Jr., his wife Carolyn Bessette Kennedy, and her sister Lauren Bessette likely were dead. The 38-year-old Kennedy was piloting a single engine Piper Saratoga carrying his wife and sister-in-law when it went down in the waters off Martha's Vineyard on Friday night (July 16) en route to his cousin Rory's wedding. John F. Kennedy Jr. was invited to Kennedy Space Center's Apollo 11 30th anniversary festivities Friday but turned down the invitation so he could attend his cousin's wedding. ["Coast Guard: JFK Jr., other likely dead," <u>Florida Today</u>, July 29, 1999, p 1A & 4A. "JFK Jr. invited to KSC," <u>Florida Today</u>, July 18, 1999, p 1A.]

◆ Hillary Clinton and the world champion U.S. women's soccer team will be here tonight, rooting for NASA's first woman commander and four other astronauts flying on shuttle Columbia. Liftoff is set for 12:36 a.m. Tuesday from Kennedy Space Center. The countdown continued smoothly Sunday as NASA officials prepared for the entourage of guests. Eileen Collins, an Air Force colonel and former test pilot who will lead Columbia's crew on a five-day mission to release an X-ray astronomy telescope. Collins, 42, was the first woman shuttle pilot chosen by NASA when she joined the agency in 1990. Last year, Collins was assigned Columbia's flight for her first job as commander. The announcement was made at the White House by President and Mrs. Clinton. The women's soccer team, fresh from a world cup victory against China on July 10, will be here, too, cheering for NASA and shuttle Columbia. Mrs. Clinton is expected to arrive in Florida after dark today, watch the liftoff, then address the KSC launch team afterward. ["Hillary, soccer team to see launch," <u>Florida Today</u>, July 19, 1999, p 1A.]

JULY 19: Space Shuttle Status Report, Monday, July 19, 1999. STS-93: The countdown for Space Shuttle Columbia's launch on mission STS-93 continues on schedule. Late stowage of the flight crew equipment into the orbiter's crew compartment concluded at about 12:30 a.m. today. Engineers completed payload inertial upper stage system checks at about 3 a.m. today and the Rotating Service Structure moved to the park position at about 7 a.m. At about 3:16 p.m. this afternoon, chilldown of the liquid propellant lines begins just prior to external tank loading. Columbia's external tank will be loaded with about 500,000 gallons of cryogenic propellant by about 6:46 p.m. Weather forecasters indicate a 30 percent probability that weather could prohibit Tuesday's launch attempt. The primary concern is the chance of coastal showers. The launch time forecast calls for clouds to be scattered at 3,000 feet and 25,000 feet; visibility at 7 miles; winds out of the Southeast at 8 knots peaking to 12 knots; temperature at 80 degrees F and a chance of coastal showers. The 24-hour and 48-hour delay forecasts improve to 20 percent chance of weather violation. Bruce Buckingham. (1999). **Kennedy Space Center Space Shuttle Status Report** [Online]. Available E-mail: domo@news.ksc.nasa.gov/subscribe shuttle-status [1999, July 19].]

◆ Judy Collins honored shuttle Commander Eileen Collins late Monday with a song commissioned by NASA through the NASA Art Program. Judy Collins performed the song, "Beyond the Sky," at the preflight briefing Monday night at the Kennedy Space Center Visitor Complex. The song was written especially for today's launch. ["New song honors Collins' mission," **Florida Today**, July 20, 1999, p 6A.]

JULY 20: Space Shuttle Status Report, Tuesday, July 20, 1999, 3:30 a.m. EDT. STS-93: Space Shuttle Columbia's July 20 launch attempt was scrubbed at the T-7 second mark in the countdown. Following a virtually flawless countdown, the orbiter's hazardous gas detection system indicated a 640 ppm concentration of hydrogen in Columbia's aft engine compartment, more than double the allowable amount. System engineers in KSC's Firing Room No. 1 noted the indication and initiated a manual cutoff of the ground launch sequencer less than one-half second before the Shuttle's three main engines would have started. Standard safing operations followed immediately. The safety of the flight crew and orbiter were not compromised at any time. The astronauts have returned to the KSC crew quarters. Following preliminary system and data evaluation, launch managers are confident that the hydrogen concentration indication was false and are proceeding with a 48-hour scrub turnaround plan. A complete review of the Shuttle's main propulsion system and related sensors is being conducted today, but managers have already determined that the hydrogen concentration was actually about 114 ppm. This measurement is within allowable limits. The launch has been rescheduled for July 22 at 12:28 a.m. Because the external ignitors at Launch Pad 39B were ignited, KSC technicians must replace them over the next two days. These ignitors burn-off the hydrogen concentration outside the orbiter, near the Shuttle main engines. The Chandra X-ray Observatory will remain powered up inside the orbiter and will not be adversely affected by the scrub. Eight middeck payloads will be removed, reserviced and installed back inside the orbiter during the down period. Weather forecasters indicate only a 10 percent chance that weather could prohibit Thursday's launch attempt. The forecast calls for scattered clouds at 3000 feet and 25,000 feet; visibility at 7 miles ; winds from the Southeast at 6 peaking to 8 knots; temperature at 80 degrees F and relative humidity at 86 percent. The only concern is the slight chance of coastal showers. Bruce Buckingham. (1999). **Kennedy Space Center Space Shuttle Status Report** [Online]. Available E-mail: domo@news.ksc.nasa.gov/subscribe shuttle-status [1999, July 20].]

◆ Space Shuttle Status Report, Tuesday, July 20, 1999, 6:51 p.m. EDT. STS-93: Following thorough evaluation of Space Shuttle Columbia's main propulsion system and hazardous gas detection system, Shuttle managers are confident in their readiness to attempt launch again on July 22 at 12:28 a.m. Engineers reviewed data last night and today that confirmed that a hydrogen leak did not cause the indication at T-8 seconds on the countdown clock. They have also confirmed that the hazardous gas detection system is working normally. Draining of liquid oxygen from the external tank was complete at 2:54 a.m. today and liquid hydrogen draining concluded at 4:03 a.m. Boil-off of residual propellants will conclude by midnight tonight. Today workers are establishing access to the external hydrogen burn-off ignitors at Launch Pad 39B and the 4-hour ignitor replacement effort begins at about 3 a.m. tomorrow. The launch team will begin reloading the external tank with 500,000 gallons of liquid propellant at about 3:38 p.m. tomorrow. Workers have removed 8 of the middeck experiments for servicing and will reinstall the secondary payloads tomorrow. The Chandra/IUS payload is ready for Thursday's launch and remains powered up in the payload bay. Weather forecasters indicate only a 10 percent chance that weather could prohibit Thursday's launch attempt. The forecast calls for few clouds at 2,500 feet and 25,000 feet; visibility at 7 miles; winds from the Southeast at 6 peaking to 8 knots; temperature at 80 degrees F and relative humidity at 86 percent. Bruce Buckingham. (1999). **Kennedy Space Center Space Shuttle Status Report** [Online]. Available E-mail: domo@news.ksc.nasa.gov/subscribe shuttle-status [1999, July 20].]

◆ With their battered and scorched Columbia lunar command module on display behind them, the crew of Apollo 11 received the nation's highest aviation honor at the National Air and Space Museum Tuesday, 30 years to the day that their mission put the first man on the moon. Mission commander Neil Armstrong and fellow astronauts Buzz Aldrin and Michael Collins each received the Smithsonian Institution's much-prized Samuel P. Langley Gold Medal from Vice president Al Gore, who as a boy was present at the launch of the Apollo 11 moon rocket from Cape Canaveral in July 1969. ["Smithsonian prize goes to Apollo 11 astronauts 30 years later," <u>The Orlando Sentinel</u>, July 21, 1999, p A-8.]

JULY 21: Space Shuttle Status report, Wednesday, July 21, 1999. STS-93: The KSC launch team continues the smooth implementation of Shuttle Columbia's 48-hour launch scrub turnaround in preparation for Thursday's 12:28 a.m. launch attempt.

The countdown clock began counting again today at 8:38 a.m. at the T-11 hour mark. Replacement of the external hydrogen burn-off ignitors at Launch Pad 39B concluded early this morning. In Firing Room No. 1, standard preflight monitoring of the Shuttle confirms that all systems are in good health and that Columbia's main propulsion system and hazardous gas detection system are ready to support launch just after midnight tonight. The Chandra payload is currently in an unpowered, flight-ready configuration, and the inertial upper stage (IUS) remains powered up. Engineers completed IUS inertial measurement unit calibration and alignment today. Refurbishment of the eight middeck experiments concluded yesterday, and the secondary payloads have been reinstalled onboard Columbia. Weather officials have improved the launch forecast to 100 percent favorable. The forecast calls for few clouds at 2,500 feet and 25,000 feet; visibility at 7 miles; winds from the Southwest at 6 knots peaking to 8 knots and temperature at 80 degrees F. Bruce Buckingham. (1999). **Kennedy Space Center Space Shuttle Status Report** [Online]. Available E-mail: domo@news.ksc.nasa.gov/subscribe shuttle-status [1999, July 21].]

◆ Brevard County might have missed the boat when it tried to attract the Boeing Co.'s new Delta 4 rocket factory, but it could pave the way for the ships that bring the rockets here to fly. The country's Public Works Department will help manage the construction of a $1 million dock at Cape Canaveral Air Station if the County Commission approves the project Tuesday at its regular meeting. Rocket components would be unloaded at the dock. The cost of the project is being picked up by a grant from the Florida Office of Tourism, Trade and Economic Development. The dock is one of the last cogs in the transportation, assembly and launching of the rockets, which are part of Boeing's entry into the U.S. Air Force's Evolved Expendable Launch Vehicle Program. Boeing has already announced plans to build a $250 million launch complex on the north side of the Cape – at the old Launch Complex 37 site – and the spaceport Flight Authority will build a $15 million, 102,000-square-foot assembly plant adjacent to it. The rockets will be shipped in smaller parts, put together and tested at the assembly plant. Boeing has a $1 billion contract with the Air Force to launch 19 of the rockets between 2002 and 2006. ["County may get chunk of Delta 4 income after all," **Florida Today**, July 22, 1999, p 1B.]

◆ Gus Grissom's Mercury capsule, Liberty Bell 7, after a long series of mechanical fits and starts, was returned to the same shore from which it blasted into space 38 years ago to the day. The capsule will be taken to the Kansas Cosmosphere and Space Center in Hutchinson, Kan., where its 25,000 pieces will be disassembled and individually cleaned. The craft will be rebuilt and after a tour of the nation, will be permanently placed on exhibit. ["Liberty Bell 7 comes home," **Florida Today**, July 22, 1999, p 1B.]

◆ First, it was a faulty sensor. This time, it was bad weather. Early today at Kennedy Space Center, NASA had to scrub shuttle Columbia's liftoff a second time when lightning drifted too close to the launch pad. The agency will give it another try early Friday. The launch window will be from 12:24 to 1:10 a.m. If the ship does not get airborne then, it won't fly until mid-August. First lady Hillary Rodham Clinton and other dignitaries – who showed up for Columbia's first launch try Tuesday – were on hand again, only to be sent home without seeing the start of NASA's 95th shuttle flight. The Tuesday launch attempt was halted seven seconds before liftoff when a faulty detector gave a bad reading indicating high levels of potentially explosive gas in the ship's engine compartment. That problem was corrected, but a stubborn thunderstorm cell 10 miles from the launch pad refused to clear out of the area in time to allow Columbia to fly. ["Shuttle scrubbed," __Florida Today__, July 22, 1999, p 1A & 2A.]

JULY 22: STS-93 Launch Weather Forecast, issued Thursday, July 22, 1999, valid Friday, July 23, 1999. Synopsis: A typical summer weather pattern will prevail for Friday morning's launch attempt. Upper level high pressure will be centered over the south Mississippi Valley. An upper level low may also be present in the central Caribbean. Disturbances will rotate from northeast to southwest over the launch area. If a disturbance is northeast of KSC near launch time, local thunderstorms are possible; if the disturbance moves west, conditions should be acceptable for launch. During the launch window of 12:24 a.m. - 1:10 a.m. EDT on Friday, July 23: Clouds: 2/8 few cumulus/stratocumulus at 3,000 feet; 3/8 scattered altocumulus at 6,000 feet; 5/8 broken cirrus at 25,000 feet; Visibility: 7 miles; Wind at Launch Pad 39-B: SW at 6-8 knots SLF: Variable at 4 knots; Temperature: 80 degrees Dewpoint: 75 degrees; Relative Humidity: 86%; Weather: Thunderstorms, rainshowers, anvil clouds; Probability of weather conditions at KSC prohibiting launch: 30%; Probability of weather conditions prohibiting tanking: 20%; Solid Rocket Booster Recovery Area: Wind: SW at 10-12 knots Seas: 3-4 feet Water temperature: 84 degrees; Sunrise: 6:38 a.m. Sunset: 8:20 p.m. Forecast by USAF45th Weather Squadron, Cape Canaveral Air Station. Bruce Buckingham. (1999). **Kennedy Space Center Space Shuttle Status Report** [Online]. Available E-mail: domo@news.ksc.nasa.gov/subscribe shuttle-status [1999, July 22].]

◆ Space Shuttle Status Report, Thursday, July 22, 1999, 4:12 a.m. STS-93: Launch managers postponed the launch of Space Shuttle Columbia today for 24 hours due to unacceptable local weather conditions. No significant technical issues were worked during the countdown and all Shuttle and payload systems remain in excellent health. The KSC launch team is executing a 24-hour scrub turnaround that will have Columbia and Chandra in a posture to launch at 12:24 a.m. Friday. Shuttle managers will convene at noon today to finalize the launch window for tomorrow's launch attempt. Weather officials are currently indicating a 30 percent probability that weather could prohibit launch on Friday. The primary threats are thunderstorms, rain showers and anvil clouds. Bruce Buckingham. (1999). **Kennedy Space Center Space Shuttle Status Report** [Online]. Available E-mail: domo@news.ksc.nasa.gov/subscribe shuttle-status [1999, July 22].]

◆ Space Shuttle Status Report, Thursday, July 22, 1999, 4:52 p.m. STS-93: Preparations are ongoing for a third launch attempt of the Space Shuttle Columbia at 12:24 a.m. Friday. Following a weather scrub early this morning, Shuttle managers implemented standard procedures for a 24-hour scrub turnaround. Columbia's external tank is being loaded for the third time with 500,000 gallons of liquid propellant. Tanking operations began at 3:15 p.m. today and will conclude at about 6:15 p.m. All Shuttle and payload systems remain in excellent health to support Friday's launch attempt. During a mission management team meeting at noon today, Chandra managers gave an updated status on Chandra's orbital power supply margins. Chandra's ability to spend an additional 20 minutes in the Earth's shadow on every orbit with no impact to mission success, adds an additional 70 minutes to the planned 46-minute window. The window for tomorrow's launch opportunity is now 1 hour and 56 minutes in duration. Mission managers still plan to deploy Chandra on flight day 1 about 7 hours and 17 minutes after liftoff. To preserve the flight crew's on-orbit sleep and work schedule, the astronauts will sleep an additional 30 minutes today before beginning their preflight preparation timeline. Each step on that timeline will begin one-half hour later tonight. The 30 minutes will be transferred from the built-in hold at T-9 minutes to the T-3 hour built-in hold. Weather officials have updated the forecast to reflect a 20 percent chance that weather could prohibit launch on Friday. The primary threats are coastal thunderstorms with debris and anvil clouds. Bruce Buckingham. (1999). **Kennedy Space Center Space Shuttle Status Report** [Online]. Available E-mail: domo@news.ksc.nasa.gov/subscribe shuttle-status [1999, July 22].]

JULY 23: Space Shuttle Status Report, Friday, July 23, 1999. STS-93: Space Shuttle Columbia, its five astronaut crew members and the Chandra X-ray Observatory lifted off from KSC's Launch Pad 39B at 12:31 a.m. today. The launch team worked no significant technical issues and weather conditions were favorable throughout the countdown. Today Columbia embarked on its 26th flight, marking the 95th launch in Shuttle history. The crew will deploy the Chandra payload about 7 hours and 17 minutes after liftoff and one hour later the inertial upper stage (IUS) solid rocket motor is slated to fire. Chandra's solar arrays will be deployed at 9:24 a.m. today and IUS separation occurs at 9:43 a.m. NASA's two solid rocket booster recovery ships Liberty Star and Freedom Star have been on station in the Atlantic Ocean since Monday. Booster recovery operations will begin this morning and managers expect the recovery ships to arrive in Port Canaveral Saturday with boosters in tow. Bruce Buckingham. (1999). **Kennedy Space Center Space Shuttle Status Report** [Online]. Available E-mail: domo@news.ksc.nasa.gov/subscribe shuttle-status [1999, July 23].]

◆ Two accidents that led to the loss or damage of multimillion dollar military satellites are being blamed on human error, investigators said Thursday. Reports showed that: An $800 million communications satellite was lost April 30 because its upper stage booster was loaded with software that had been incorrectly written by computer engineers. A $45 million military navigation satellite was drenched at its Cape Canaveral launch pad May 8 after a heavy thunderstorm swept through the central Florida area, causing at least $2.1 million in damage to the spacecraft. ["Satellite accidents due to human error, officials say," Florida Today, July 23, 1999, p 2A.]

JULY 25: A potentially dangerous fuel leak in one of shuttle Columbia's main engine nozzles could explain why the spaceship wound up in an orbit seven miles lower than planned after liftoff Friday. Pictures taken of the shuttle in flight suggest Columbia's right nozzle began losing liquid hydrogen a second or two before launch and continued to do so for the entire 8 ½ minute ascent. One or two cracked coolant lines may have leaked as much as 5 pounds of hydrogen per second. There have been smaller leaks on previous flights. But NASA still is not certain a leak occurred on this one. During the climb to orbit, Columbia's three main engines cut off about a second earlier than planned. An analysis found the shuttle was short 4,000 pounds of liquid oxygen, used along with hydrogen to power the engines. Engineers traced the oxygen shortage to the possible hydrogen leak. Each of the bell-shaped engine nozzles, which stand about 9 feet tall, is made up of 1,080 stainless steel tubes about a ¼-inch in diameter. The tubes circulate super-cold liquid hydrogen from the shuttle's external tank to the combustion chamber above the nozzle. This process keeps the nozzle cool and preheats some of the fuel for more efficient burning. Mission managers theorize that as hydrogen leaked from one or two of the tubes, the engine's computer compensated by providing more thrust, depleting the supply of liquid oxygen faster than expected. Another, unrelated, launch problem may be closer to a quick fix. Less than 10 seconds after liftoff, computers controlling two of the shuttle's main engines were knocked out by a short circuit. Backup computers took over, and Columbia continued to fly as normal. But had one of the backups been lost, an engine would have shut down. While inspecting circuit breakers for the second time, Collins discovered a tripped breaker than had been overlooked before. Flight controllers have narrowed the list of possible causes. ["Shuttle may have leaked hydrogen fuel during launch," The Orlando Sentinel, July 26, 1999, p A-7.]

◆ A Delta II rocket lifted off at 3:46 a.m. Sunday, carrying four Globalstar satellites from Cape Canaveral Air Station. The launch was delayed one day to give shuttle Columbia another attempt at liftoff. ["Delta II," Florida Today, July 26, 1999, p 1A.]

JULY 26: STS-93 Landing Weather Forecast, Tuesday, July 27, 1999. Date issued: July 26, 1999, 10 a.m. EDT. Synopsis: Southwesterly low-level winds will slow the onset of the sea breeze and limit its inland penetration. An approaching weak upper level trough will help trigger showers and thunderstorms along the sea breeze tomorrow afternoon. The upper level system is forecast to be just south of KSC by tomorrow evening. Passage of the upper level system and sunset will combine to decrease thunderstorm coverage over Florida during the evening hours. There is only a slight chance that a thunderstorm or shower will be within 30 miles of the Shuttle Landing Facility for the first landing opportunity. Little change in the forecast is called for on Wednesday night. Weather at Edwards Air Force Base is expected to remain hot and breezy in the afternoon with winds slowly subsiding by possible landing times Tuesday and Wednesday nights. However, due to the generally favorable weather forecast at KSC, Dryden will not called up for landing support on Tuesday. Kennedy Space Center Shuttle Landing Facility: Valid: 11:20 EDT; Clouds: FEW 3,000 FEW 25,000 (1/8 - 2/8 sky coverage); Visibility: 7 miles; Wind: SW/3 knots, peak 5 knots; Temperature: 80; Dewpoint: 74; Relative Humidity: 82%; Precipitation: Slight chance of thunderstorms within 30 miles of KSC located mainly to the west of the SLF at the first landing opportunity. Bruce Buckingham. (1999). **Kennedy Space Center Space Shuttle Status Report** [Online]. Available E-mail: domo@news.ksc.nasa.gov/subscribe shuttle-status [1999, July 26].]

◆ After years of nibbling away at NASA's budget, a key House committee voted Monday to take a huge bit, cutting almost $1.4 billion from the space agency. The House Appropriations subcommittee for veterans, housing and other federal agencies cut NASA's budget as part of a larger effort to stay within tight budget limits. Democrats accused Republicans of hewing, despite the fact the nation is running a surplus, in order to fund their tax-cut plans. Monday's vote would shave the combined budgets of various agencies from $71.9 billion in the current year to $70.5 billion for the year that begins Oct. 1. NASA bore almost all the brunt, with a cut of about 10 percent. In the past five years, NASA's budget has dropped from about $14.5 billion to $13.66 billion. The subcommittee voted to cut NASA to $12.3 billion. ["NASA could face another astronomical budget cut," **The Orlando Sentinel**, July 27, 1999, p A-1 & A-4.]

◆ An asteroid targeted for a close encounter by a NASA spacecraft this week was renamed Monday to honor Louis Braille, the inventor of the raised-dot alphabet that enables blind people to read. Kerry Babcock, 38, of Port Orange submitted the winning entry in naming the contest sponsored by The Planetary Society, a Pasadena group that promotes interest in space exploration and education. The Kennedy Space Center software engineer said he has been interested in Braille's work for years and even named his daughter after the French inventor. The asteroid was formerly known as 1992KD. ["Spacecraft's target asteroid named after Braille," **Florida Today**, July 27, 1999, p 6A.]

JULY 28: Space Shuttle Status Report, Wednesday, July 28, 1999. STS-93: Space Shuttle Columbia and its five astronaut crew members successfully landed late last night on KSC's Shuttle Landing Facility runway 33, completing its 5 day, 1.8 million mile journey. July 27, 1999 Eastern daylight landing time events are as follow: Main Gear Touchdown: 11:20:37 p.m. EDT (MET: 4 days, 22 hours, 49 minutes, 37 seconds); Nose Gear Touchdown: 11:20:44 p.m. EDT (MET: 4 days, 22 hours, 49 minutes, 44 seconds); Wheels Stop: 11:21:22 p.m. EDT (MET: 4 days, 22 hours, 50 minutes, 22 seconds); This was the 19th consecutive Shuttle landing at the Florida spaceport and the 12th night landing in Shuttle program history. Following an early morning press conference, the crew of STS-93 flew back to their homes and families in Houston, TX. Preliminary indications of the orbiter after touchdown show its lower surface sustained 155 total hits, of which 40 had a major dimension of 1-inch or larger. Further assessments will continue today. The main landing gear tires were reported to be in good condition for a landing on the KSC concrete runway. Several hours after touchdown, Columbia was towed to orbiter Processing Facility bay 3 where post mission inspections began. Today, workers are off loading the unused onboard cryogenics and safing the vehicle for additional inspections. Tonight, engineers are expected to gain access to the orbiter's main engines. They will then begin preliminary inspections of what early indications show to be a possible small hydrogen leak that occurred during ascent on the No. 3 engine nozzle. A possible source of the leak is from small tubes that circulate hydrogen around the nozzle, cooling the nozzle and conditioning the hydrogen before it is burned. Bruce Buckingham. (1999). **Kennedy Space Center Space Shuttle Status Report** [Online]. Available E-mail: domo@news.ksc.nasa.gov/subscribe shuttle-status [1999, July 28].]

◆ Eileen Collins kept calm in the driver's seat when shuttle Columbia had some problems during liftoff. Then she ran a smooth mission in orbit to unleash a powerful new telescope on the universe. Collins flew her spaceship back to Florida in a return so gentle one of her crew called it a "kiss landing." NASA's first woman shuttle commander racked up respect and kudos Wednesday, when Columbia's crew left Kennedy Space Center for a welcome back party in Houston with Vice President Al Gore. ["Consensus of mission: 'Eileen rocks'," Florida Today, July 29, 1999, p 1B.]

◆ With the shuttle Columbia back at Kennedy Space Center, NASA confirmed its suspicions Wednesday that one of the craft's engine nozzles leaked fuel during last week's liftoff. Engineers began investigating the leak and an unrelated short circuit in Columbia's computers shortly after the shuttle landed late Tuesday. They hope to figure out by early next week what caused both problems. The effect on future shuttle flights remains to be seen. Preparations for the next mission – a Sept. 16 radar-mapping mission aboard Endeavour – are moving ahead. ["NASA confirms leak on engine's nozzle," The Orlando Sentinel, July 29, 1999, p A-11.]

JULY 29: Space Shuttle Status Report, Thursday, July 29, 1999. STS-93: Following Columbia's tow to the Orbiter Processing Facility bay 3 early Wednesday morning at the conclusion of its STS-93 mission, workers continue safing operations and post-mission inspections. Columbia is being readied for ferry to Palmdale, CA at the end of September for an extended period of structural inspections and orbiter modifications. Last night, engineers in the OPF made initial visual inspections of the No. 3 main engine nozzle and the apparent hotwall ruptures in three adjacent coolant tubes. Engineers believe these ruptures resulted in a small hydrogen leak that occurred during Columbia's launch last week. Overnight, the damaged area was removed from the nozzle and sent to the Rocketdyne facility in Conoga Park, CA for analysis. Also, access to the orbiter's aft engine compartment continues in order to allow workers to troubleshoot a problem that caused an apparent short circuit on one of the electrical busses that feed controllers on the right and center main engines. The center main engine primary controller was shutdown shortly after booster ignition and the backup controller for the right main engine was disqualified. The solid rocket boosters were towed to Hangar AF and inspections begin July 26. Initial indications show both boosters to be in excellent condition following launch. Bruce Buckingham. (1999). **Kennedy Space Center Space Shuttle Status Report** [Online]. Available E-mail: domo@news.ksc.nasa.gov/subscribe shuttle-status [1999, July 29].]

JULY 30: A gold-plated engine plug the size of a small nail shook loose during shuttle Columbia's recent launch, apparently triggering a hydrogen fuel leak that could have led to an emergency landing. That was the word Friday from NASA investigators who are trying to determine how coolant lines in a shuttle engine nozzle burst during launch, leaving Columbia short of its intended orbit. Columbia and five astronauts – including Eileen Collins, the nation's first woman space commander – lifted off from Kennedy Space Center on July 23 on a successful mission to deploy a $1.5 billion astronomical observatory. Three coolant loops inside a shuttle engine nozzle, however, ruptured seconds before liftoff, creating a 5-pound-a-second liquid hydrogen leak that continued throughout Columbia's 8 ½-minute climb into orbit. NASA officials said the leak could not have led to an explosion in flight. Soon after Columbia landed late Tuesday, engineers discovered three ragged holes, each the size of a fingernail, in cooling lines within the nozzle of one of the ship's three main engines. Sections of the ruptured lines were removed from the leaky nozzle and sent to a California engine factory for analysis. Using a scanning electronic microscope to magnify the holes, engineers discovered tiny dents at the leading edge of each puncture – a strong indication the bell-shaped nozzle was hit by debris when Columbia's engines ignited. The debris likely was a small tapered plug that apparently shot through the engine's combustion chamber at about four times the speed of sound – 3,000 mph. Made of the metal alloy Inconel, the plug traveled 6 to 8 feet before smashing into the inside of the nozzle. The plug probably melted in a river of flame after the shuttle's engines ignited. The plug became NASA's leading suspect after engineers discovered it missing from the engine's main injector. Columbia also sustained a momentary short circuit five seconds after liftoff, which knocked out the prime computers for two of its three main engines. Technicians began inspecting electrical wiring within the shuttle orbiter Friday in a bid to pinpoint the source of the short. The inspections are expected to take several days. ["Loose engine plug likely caused shuttle fuel leak," **Florida Today**, July 31, 1999, p 1A & 4A.]

DURING JULY: NASA and Boeing are investigating why technicians here misaligned the 3,000-lb. FUSE (Far Ultraviolet Spectroscopic Explorer) when placing it atop its Delta II booster prior to the observatory's launch June 24. The spacecraft was launched safely and is operating normally. But the incident, coming in the wake of a series of launch failures, is another example of how human error can affect space operations. The 5.3-deg. Misalignment with the Delta's second stage, discovered several days before launch, forced an exhaustive government/contractor assessment to determine whether it posed any risk to the $215-million mission. Boeing said the reviews took place at both its Huntington Beach Calif., and Cape facilities. ["Cape

Technicians Misalign Satellite," <u>Aviation Week & Space Technology</u>, July 5, 1999, p 23.]

◆   A NASA investigative board has determined that the loss of the agency's $73-million Wide-Field Infrared Explorer (Wire) shortly after launch on Mar. 4 was caused by an improperly designed electronic component that triggered a premature firing of pyrotechnics. Electrical power almost instantly reached the pyrotechnics, ejecting the telescope's cover, as controllers initially had suspected. As a result, frozen hydrogen used to cool the Wire's infrared detectors was exposed to sunlight, converted to gas and vented, rendering the telescope unusable. ["Design Flaw," <u>Aviation Week & Space Technology</u>, July 12, 1999, p 13.]

AUGUST

AUGUST 2: Boeing workers Wednesday will replace one of the solid rocket boosters attached to a Delta 2 rocket set for launch later this month, the company said Monday. Officials decided to make the precautionary replacement after discovering the original booster had "superficial scratches" on its outer casing, said Boeing spokeswoman Erin Lutz. The January 1997 mid-air explosion of a Delta 2 rocket was caused when a solid rocket motor casing ruptured. Investigators determined the casing was damaged during pre-flight processing. As a result of the extra work, the upcoming launch has been delayed two days to Aug. 17. The three-minute window will open at 12:37 a.m. ["Delta 2 rocket booster will be replaced," <u>Florida Today</u>, August 3, 1999, p 1B.]

AUGUST 4: After its arrival at the Shuttle Landing Facility, the Multi-Purpose Logistics Module Raffaello was transported to the Space Station Processing Facility (SSPF). The SSPF now holds two MPLMs, the other being Leonardo. The two MPLMs are scheduled to launch on back-to-back missions: Leonardo on June 29, 2000, and Raffaello on July 27, 2000. ["Raffaello moved into SSPF to begin testing," **KSC Countdown**, August 10, 1999.]

AUGUST 6: On the last mission to the new international space station, several astronauts suffered nausea, headaches and other symptoms that have NASA scrambling for solutions to a suspected air-quality problem before another crew visits the growing outpost. Moreover, the crew of the shuttle Discovery did not report the difficulties until debriefings two weeks after their landing, a delay that NASA authorities acknowledge has hampered the search for possible solutions. The seven-person crew returned from the 10-day supply mission on June 6. NASA officials say they think unexpected buildups of carbon dioxide, fumes from chemicals and restricted air circulation may have been the sources of the problem. Frank Culbertson, NASA's space station operations manager at the Johnson Space Center in Houston, said the problems were not serious but the agency was taking steps to see that they did not recur. ["Space illness puzzles NASA," <u>The Orlando Sentinel</u>, August 7, 1999, p A-3.]

AUGUST 9: A wiring defect that escaped detection prior to shuttle Columbia's recent launch caused an electrical short that left the crew one failure away from an unprecedented emergency landing, officials said Monday. With the shuttle back in its Kennedy Space Center hangar, technicians traced the short circuit to an exposed wire hidden deep within the ship's 122-foot-long fuselage. "They were able to pinpoint it to a wire that had worn insulation," said KSC spokesman Bruce Buckingham. Technicians also found evidence of electrical arcing between the exposed wire and a metal screw head, which apparently wore through the insulation over 25 previous flights of Columbia, NASA's oldest orbiter. Coming five seconds after a July 23 liftoff, the short circuit caused the main computers on two of Columbia's three liquid-fueled engines to crash. Had backups to either computer failed, Columbia and its crew would have been forced to make a risky landing at KSC or an emergency runway in Africa – something never attempted in 95 shuttle flights since 1981. Buckingham said wire harness inspections will continue to be done on Columbia as well as shuttle Atlantis. However, they will not be done on shuttle Endeavour, which is scheduled for liftoff Sept. 16 on an 11-day radar mission to map Earth's surface. The reason is the radar equipment is already packed in the ship's cargo bay, and would have to be removed for workers to gain access to the area where the wires are located. Engineers are less concerned about Endeavour because it is the youngest ship in the fleet, and its harnesses have not been subjected to as much wear as Columbia's. The short circuit was one of two serious problems that cropped up during Columbia's 8 ½-minute climb into orbit. One of its engines started leaking potentially explosive hydrogen fuel after three coolant tubes within its nozzle were ruptured. That problem was traced to a small plug which came loose during engine ignition. The plug smashed into the lines, weakening them so much that they burst from internal pressure. Columbia's crew – led by Eileen Collins, the nation's first female shuttle commander – successfully deployed a $1.5 billion astronomical observatory during a five-day flight. ["NASA: Wiring flaw caused electrical short on shuttle," <u>Florida Today</u>, August 10, 1999, p 1A.]

AUGUST 10: Space Shuttle Status Report, Tuesday, August 10, 1999. STS-99: Endeavour is now scheduled to roll to the VAB no earlier than Thursday morning. This afternoon Shuttle managers reconvened to review the results of electrical wiring inspections on orbiters Columbia and Atlantis. Also presented in the meeting were the results of laboratory analysis on Columbia's damaged electrical wire. The analysis revealed that an isolated incident of mechanical or worker contact with the wire caused the wire to press against a marred screw head. The subsequent wire damage lead to the electrical short during Columbia's STS-93 launch. Since the laboratory results were not available until today, engineers will review the new information and bring a recommendation to the Shuttle program tomorrow at midday. Tomorrow afternoon, Shuttle managers will decide if orbiter Endeavour is ready for Thursday's planned rollout activities. STS-103: Technicians completed servicing of Discovery's onboard ammonia system. Following successful main engine installation last week, engine securing efforts are under way today. The payload premate test is also ongoing. Later this week, workers will install the orbiter's waste collection system and install the external airlock hatch. STS-101: Orbiter Atlantis is being temporarily stored in VAB high bay 2, awaiting the departure of Shuttle Endeavour from OPF bay 2. Bruce Buckingham. (1999). **Kennedy Space Center Space Shuttle Status Report** [Online]. Available E-mail: domo@news.ksc.nasa.gov/subscribe shuttle-status [1999, August 10].]

AUGUST 11: Space Shuttle Status Report, Wednesday, August 11, 1999. STS-99: Endeavour is scheduled to roll to the Vehicle Assembly Building no earlier than Friday morning. Postponing Endeavour's transfer to the VAB from today to Friday will not impact the Sept. 16 target launch date for mission STS-99. Shuttle engineers continue to review the data gathered during recent wiring inspections of orbiters Columbia and Atlantis. Also under evaluation are the results from yesterday's laboratory analysis performed on Columbia's damaged wire in KSC's malfunction laboratory. Tomorrow at midday Shuttle managers will determine Endeavour's readiness to roll to the VAB. STS-103: Technicians completed testing of Discovery's power reaction and storage distribution system. Shuttle main engine securing efforts and payload premate testing continue. Later this week, workers will install the orbiter's waste collection system and install the external airlock hatch. STS-101: Orbiter Atlantis is being temporarily stored in VAB high bay 2, awaiting the departure of Shuttle Endeavour from OPF bay 2. Bruce Buckingham. (1999). **Kennedy Space Center Space Shuttle Status Report** [Online]. Available E-mail: domo@news.ksc.nasa.gov/subscribe shuttle-status [1999, August 11].]

◆ A work place accident – rather than wear and tear – likely caused a wiring defect that triggered a serious electrical short on shuttle Columbia's recent flight, NASA officials said Wednesday. A new lab analysis shows that a worker probably stepped on a wire within the shuttle's fuselage, pressing it up against the marred head of a screw. Insulation was inadvertently nicked as a result, leaving the wire exposed, officials said. Apparently, strong launch vibrations then pushed the exposed wire against the metal screw head, triggering a short circuit five seconds after a July 23 liftoff. "We feel like it was a single isolated incident where a worker made contact with the wire, which then made contract with the screw and nicked the insulation," said Joel Wells, a spokesman at NASA's Kennedy Space Center. The electrical short crashed the main computers on two of Columbia's three liquid-fueled main engines. Had back-ups to either computer failed, Columbia's crew – led by Eileen Collins, NASA's first female shuttle commander – would have been forced to make an unprecedented emergency landing attempt at KSC or a runway in Africa. Earlier this week, NASA officials believed the defect was wear and tear the wire had been subjected to during 25 previous flights of Columbia, NASA oldest orbiter. The new analysis has prompted another round of extra inspections to wiring on both Columbia and Atlantis. The inspections are intended to give managers an idea of whether wiring defects might exist on shuttle Endeavour, NASA's newest orbiter. Plans to move Endeavour from its hangar to the Vehicle Assembly Building at KSC have been delayed until at least Friday. ["Accident, not wear, linked to shuttle short circuit," <u>Florida Today</u>, August 12, 1999, p 9A.]

AUGUST 12: Space Shuttle Status Report, Thursday, August 12, 1999. STS-99: Shuttle managers today decided to delay the rollout of Shuttle Endeavour from OPF bay 2 to the Vehicle Assembly Building to conduct extensive wiring inspections and preventative wire maintenance in the orbiter's payload bay. In depth evaluation of payload bay wiring aboard orbiters Columbia and Atlantis revealed the potential for damaged wire to exist in Endeavour's payload bay. The additional work will delay the STS-99 launch to at least early October. Tomorrow, workers will begin preparations to remove the SRTM payload from Endeavour's payload bay to gain access to the lower cable trays that run the length of the orbiter's midbody. Once access is established, Shuttle engineers and technicians will begin necessary inspection and mitigation efforts. The impact of this delay and the unplanned wiring work needed on the rest of the Shuttle fleet is being assessed. STS-103: Technicians completed testing of

Discovery's power reaction and storage distribution system. Shuttle main engine securing efforts and payload premate testing continue. Later this week, workers will install the orbiter's waste collection system and install the external airlock hatch. STS-101: Orbiter Atlantis is being temporarily stored in VAB high bay 2, awaiting the departure of Shuttle Endeavour from OPF bay 2. Bruce Buckingham. (1999). **Kennedy Space Center Space Shuttle Status Report** [Online]. Available E-mail: domo@news.ksc.nasa.gov/subscribe shuttle-status [1999, August 12].]

◆ The payload flight hardware for the Third Hubble Space Telescope Servicing Mission (SM-3A) arrived today at Kennedy Space Center aboard a C-5 air cargo plane. It was shipped from NASA's Goddard Space Flight Center in Greenbelt, MD. After off loading from the C-5, the shipping container was taken to the Payload Hazardous Servicing Facility (PHSF) located in the KSC Industrial Area. There final integration of the payload elements will occur and each will be fully tested. The Third Servicing Mission (SM-3A) is a "call-up" mission which is being planned due to the need to replace portions of the spacecraft's pointing system, the gyros, which have begun to fail. Only three of its six gyroscopes – which allow the telescope to point at stars, galaxies and planets – are working properly. The telescope needs at least three gyroscopes to operate properly. ["Payload for Third Hubble Servicing Mission arrives at KSC," **NASA News Release #69-99,** August 12, 1999.]

◆ NASA will delay its next shuttle mission to make sure the type of serious electrical short encountered on its most recent flight won't endanger future astronaut crews. In a bid to avoid another close call in flight, NASA on Thursday ordered fleet wide inspections of suspect wiring – a move that will push back the planned Sept. 16 launch of shuttle Endeavour to at least Oct.7. A high-priority Hubble Space Telescope repair mission slated for launch Oct. 14 probably won't fly until late October or early November. NASA officials said they won't risk a repeat of an electrical short circuit that left five astronauts one failure away from an emergency landing on shuttle Columbia's July 23 flight. "This is the prudent thing to do under the circumstances," said senior shuttle manager Don McMonagle, a former astronaut. An Oct. 7 launch for Endeavour's 11-day flight would push back the Hubble mission to at least late October. The reason: NASA engineers need about nine days between a shuttle landing and the next launch to review data from the previous flight. ["Next shuttle won't go up until October," Florida Today, August 13, 1999, p 1A.]

AUGUST 13: Space Shuttle Status Report, Friday, August 13, 1999. STS-99: Following yesterday's decision to holdup Shuttle Endeavour's transfer to the Vehicle Assembly Building, workers moved the orbiter transport system out of Orbiter Processing Facility bay 2 today. Technicians will gain access to the orbiter's crew module today and begin preparations to open the payload bay doors tomorrow. Shuttle engineers and technicians will meet tomorrow to develop a wiring inspection and preventative maintenance plan to be applied across the Shuttle fleet. Managers plan to begin removing the SRTM payload from the cargo bay late next week. Once the payload is removed, technicians will begin Endeavour's midbody wiring work. The additional work will delay the STS-99 launch to at least early October. STS-103: Shuttle managers have decided to employ the same wiring inspections and maintenance that are being planned for Endeavour across the Shuttle fleet. Since Discovery's payload bay is already accessible, maintenance efforts will likely begin early next week. Technicians completed Shuttle main engine securing efforts this week. Main propulsion system and main engine leak checks are ongoing. Potable water servicing and external airlock hatch installation are in work. Forward reaction control system installation remains on schedule. STS-101: Orbiter Atlantis is being temporarily stored in VAB high bay 2, awaiting the departure of Shuttle Endeavour from OPF bay 2. Wiring inspections have been in work all week in Atlantis and maintenance efforts will begin in the days ahead. Bruce Buckingham. (1999). **Kennedy Space Center Space Shuttle Status Report** [Online]. Available E-mail: domo@news.ksc.nasa.gov/subscribe shuttle-status [1999, August 13].]

◆ Demolition crews next month will blow up a Cape Canaveral launch pad that served as the starting point for historic NASA missions to explore other planets. Lockheed Martin engineers have decided to demolish Launch Complex 41's towering metal structures and transform the 36-year-old Titan rocket pad into a launch site for new Atlas 5 rockets. The event is a first for the Space Coast. Never before has a commercial company undertaken such an operation at the air station. On Sept. 12, the pad's launch tower and rolling service structure will be toppled to the ground in less than a minute by explosive charges, said Adrian Laffitte, Lockheed Martin's Atlas program director. The huge steel structures won't be needed for the company's new fleet of rockets, which will begin flying from the base in late 2001. The reason: The Atlas 5s will be assembled and readied for launch in a nearby building, then rolled to the open site on a mobile transporter 24 hours before liftoff. With such a short wait on the pad, the new rockets won't require structures like those in place at all other Cape Canaveral launch pads. Lockheed Martin selected the explosive method after its construction contractor suggested the plan to save time and save money. The less dramatic option of dismantling the towers piece by piece would have taken about three months, Laffitte said. Built in the early 1960s, Complex 41 saw its first launch Dec. 21, 1965. After that, it was used to launch NASA's Viking probes to Mars in 1975 and the two Voyager missions in 1977 that toured the solar system's outer planets. The pad was refurbished to launch the Air Force's Titan 4 rockets, America's most powerful unmanned boosters. "There is a lot of history at Complex 41," said Brig. Gen. Randall Starbuck, 45th Space Wing commander. ["Explosives will topple historic pad," <u>Florida Today</u>, August 14, 1999, p 1B.]

AUGUST 15: The grand finale for this summer's record-setting string of launches by Boeing's Delta 2 rocket fleet is set for just after midnight. If all goes as planned, the company will launch another Delta 2 rocket, carrying four Globalstar cellular telephone satellites, at 12:37a.m. Tuesday from Cape Canaveral Air Station. The launch team will have 3-minute window to get the mission airborne. The $110 million flight is the fifth and final launch scheduled during an intense 68-day period that began June 10. The cluster of launches has produced major strides in assembling Globalstar's space based satellite constellation and has delivered a NASA telescope into Earth orbit. With this last launch, if successful, Boeing will have placed 17 satellites into space in a record pace of just more than two months. The launch surge began when Globalstar ordered additional Delta 2 rockets to deploy half of the international consortium's planned 48-satellite system. ["Launch will cap record Delta run," **Florida Today**, August 16, 1999, p 1B.]

AUGUST 19: Space Shuttle Status Report, Thursday, August 19, 1999. STS-99: Yesterday, workers removed the Shuttle Radar Topography Mission payload from Endeavour's payload bay and today technicians are establishing access to the orbiter's midbody for planned wiring inspections. Inspections are expected to begin this weekend. With wiring inspection and maintenance plans now in place, implementation efforts are in progress across the Shuttle fleet. Shuttle managers are reviewing several manifest options this week and could establish the new target launch dates for 1999 as early as next week. Engineers must first define the time that will be needed to complete the fleet-wide wiring maintenance effort. Shuttle Endeavour currently remains slated for launch in early October. STS-103: Wiring inspections are about 25 percent complete aboard Shuttle Discovery. Earlier access allowed workers to begin inspections on Discovery this week. Technicians have completed installation of Discovery's external airlock hatch. Electrical connections are complete for the orbiter's forward reaction control system. Main engine heatshield installation is in work. STS-101: Orbiter Atlantis is currently being stored inside Vehicle Assembly Building high bay 2, awaiting the departure of Shuttle Endeavour from OPF bay 2. Wiring inspections in Atlantis' payload bay are complete and wireless video modifications have resumed. Atlantis will be moved to the VAB transfer aisle next week to accommodate external tank transfer activities. Space Shuttle Columbia/OV-102: Preparation for Orbiter Maintenance Down Period (OMDP): Orbiter Columbia is undergoing routine post flight deservicing in Orbiter Processing Facility bay 3. Also, workers are preparing the spacecraft for its upcoming Orbiter Maintenance Down Period (OMDP) in Palmdale, CA. Columbia is scheduled to be mounted atop NASA's modified Boeing 747 on Sept. 22. The ferry flight is scheduled to begin on Sept. 23 with an overnight stop at Luke Air Force Base in Arizona. Current plans have Columbia arriving in Palmdale, CA on Sept. 24. Because the orbiter can not be flown through precipitation of any kind, ferry flight plans are contingent upon weather conditions in the flight path. Bruce Buckingham. (1999). **Kennedy Space Center Space Shuttle Status Report** [Online]. Available E-mail: domo@news.ksc.nasa.gov/subscribe shuttle-status [1999, August 19].]

AUGUST 24: Beginning a yearlong celebration to mark 50 years of rocket launches from Cape Canaveral Air Station, Lt. Gov. Frank Brogan on Tuesday pledged that Florida will remain a key state for future space flights. Brogan, who is Gov. Jeb Bush's liaison for space-related issues, said government leaders, the business sector and the public cooperated to build the state's reputation as the No. 1 location for space launches. That collaboration could be the key to continued success, he said. "This legacy, this history, is really the platform for what Florida has the potential to be in the 21st century, if we partner," Brogan said. Brogan addressed more than 250 people at the gathering sponsored by the Florida Space Business Roundtable. It was billed as the kickoff for numerous events that will lead to July 24, the 50th anniversary of the first successful launch of the Bumper rocket from Cape Canaveral. The Bumper was a rocket consisting of a German V-2 missile with a U.S. Army WAC Corporal rocket as the second stage. The first six Bumpers were launched from White Sands Proving Grounds in new Mexico. The final two launches lifted off from Cape Canaveral. As important as the Bumper program was to the development of more advanced rockets, including those used by NASA, there has been nothing to note that early launch at the Air Station. To make up for that oversight, Brig. Gen. Donald P. Pettit, the new commander of the 45th Space Wing at Patrick Air Force Base, said: "(We are) marking the spot where all this began." At that prompting, Norris Gray, an original member of the Bumper program team, unveiled a street sign reading "Bumper Road" marking the location of the launch. Also unveiled was a commemorative logo marking the Cape rocket launches. ["Brogan ignites year of launch celebrations," Florida Today, August 25, 1999, p 1B.]

◆ NASA's next trip to the International Space Station most likely will be delayed until early next year because of fleetwide inspections to shuttle wiring systems, officials said Tuesday. A mission to repair the Hubble Space Telescope, meanwhile, might move up in NASA's shuttle launch queue, delaying an already postponed radar-mapping flight another month. The anticipated changes to NASA's shuttle launch schedule are expected to be completed late this week or early next week. "It is a direct result of the fact that we are doing the wiring inspections and repairs," said James Hartsfield, a spokesman for NASA's Johnson Space Center in Houston. Fleetwide inspections were ordered after an electrical short circuit caused two engine computers to fail five seconds into Columbia's July 23 flight. Technicians traced the short to a defective wire hidden within the shuttle's fuselage. The subsequent discovery of another damaged wire on Columbia prompted inspections on all shuttles. Since then, similar damage has been found on Endeavour, Atlantis and Discovery. Once the work is completed, NASA will lay out a new schedule. ["Wiring checks zap shuttle schedule," Florida Today, August 25, 1999, p 1A.]

AUGUST 25: A team of KSC experts has developed the complex umbilical system for the X-33, a half-scale prototype of the planned reusable launch vehicle dubbed VentureStar. [KSC Countdown, August 26, 1999.]

AUGUST 26: Space Shuttle Status Report, Thursday, August 26, 1999. STS-99: Following the removal of the Shuttle Radar Topography Mission payload from Endeavour's payload bay last week, technicians are continuing with inspections and repairs as necessary of the wiring in the orbiter's midbody and aft engine compartment. These inspections were required for all Shuttle orbiters following mission STS-93 when a damaged wire to a main engine controller shorted out. Preventative measures are under way to preclude similar wiring damage on all vehicles. At this time, over 70 percent of the work on Endeavour is complete. The SRTM payload is currently in the Vertical Processing Facility and will remain there until all wiring work on Endeavour is complete. Launch options are being reviewed this week for Endeavour as well as Discovery. No specific target launch dates are expected to be announced until later next week. STS-103: Wiring inspections are about 50 percent complete aboard the orbiter Discovery. Launch options are being reviewed this week for Endeavour as well as Discovery. No specific target launch dates are expected to be announced until later next week. STS-101: Orbiter Atlantis is currently being stored inside the Vehicle Assembly Building, awaiting the opening of an Orbiter Processing Facility bay. Wiring inspections in Atlantis' payload bay are complete. Atlantis was moved yesterday to the VAB transfer aisle from high bay 2 to accommodate external tank transfer activities. Shuttle managers today approved the delay of mission STS-101 to Jan. 22, 2000. Space Shuttle Columbia/OV-102: Preparation for Orbiter Maintenance Down Period (OMDP): Orbiter Columbia continues to undergo routine post flight-deservicing in Orbiter Processing Facility bay 3. Wiring inspections are complete and the wire that experienced the short on mission STS-93 has been repaired. Preventative measures have been taken to preclude similar damage. Workers are preparing the orbiter for its upcoming Orbiter Maintenance Down Period (OMDP) in Palmdale, CA. Columbia is scheduled to be mounted atop NASA's modified Boeing 747 on Sept. 22. The ferry flight from KSC is scheduled to begin on Sept. 23 with an overnight stop at Luke Air Force Base in Arizona. Current plans have Columbia arriving in Palmdale on Sept. 24. Because the orbiter can not be flown through precipitation of any kind, ferry flight plans are contingent upon weather conditions in the flight path. Bruce Buckingham. (1999). **Kennedy Space Center Space Shuttle Status Report** [Online]. Available E-mail: domo@news.ksc.nasa.gov/subscribe shuttle-status [1999, August 26].]

AUGUST 27: A new organization meant to boost Florida-based space research will meet next month to decide its "structure, programs and priorities." The organization, the Florida Space Research Institute, initially will help administer $200,000 in Spaceport Florida Authority funds, including matching them with NASA grants to Florida universities and colleges. The institute's seven-member board includes representatives from the Spaceport Florida Authority, Enterprise Florida Inc., the Florida Aviation Aerospace Alliance and the Florida Space Business Roundtable and Kennedy Space Center contractors, including Bionetics, Boeing Co. and Command & Control Technologies. The institute also will support a partnership between Kennedy Space Center and Florida universities to develop a KSC-based biotechnology and microgravity payload facility that supports research projects for the space shuttle, the Space Station and Lockheed martin's proposed VentureStar reusable launch vehicle. ["Group promotes Fla. Space research," <u>Florida Today</u>, August 28, 1999, p 1C.]

◆   Further inspections of NASA's shuttle fleet have found wire damage in all four of the agency's orbiters. Future flights were put on hold and inspections ordered after a short circuit knocked out a pair of main-engine computers seconds after shuttle Columbia's July 23 liftoff. After the mission, investigators found a wire had been pressed against the head of a damaged screw during preflight preparations. So far, 20 additional sections of damaged wiring have been found in shuttle Endeavour – the next orbiter scheduled to fly – including evidence of an old short circuit. Nine damage areas have been discovered on Discovery. Inspections on those two ships are continuing. Inspectors are done checking Columbia and shuttle Atlantis. Atlantis had one area of damage; Columbia had two. Shuttle managers are expected next week to announce new launch dates for Endeavour's radar-mapping mission and a flight by Discovery to repair the Hubble Space Telescope. Which mission will go first has not been determined, but neither is likely to launch before mid-October. NASA already has moved a space-station supply flight aboard Atlantis from December to January. Columbia is scheduled for transport to Southern California next month for a year of refurbishing. ["All 4 space shuttles have damaged wires," **The Orlando Sentinel**, August 28, 1999, p A-3.]

AUGUST 31: Space Shuttle Status Report, Tuesday, August 31, 1999. STS-99: Endeavour's payload bay doors have been reopened following their closure last Friday due to the potential for tropical storm strength winds from Hurricane Dennis over the weekend. Technicians this week will continue with inspections and repairs as necessary of the wiring in the orbiter's midbody. These inspections were required for all Shuttle orbiters following mission STS-93 when a damaged wire to a main engine controller shorted out.   Preventative measures are under way to preclude similar wiring damage on all vehicles.   Engineers are continuing to analyze a bent freon line associated with payload bay support equipment for the SRTM payload. The bent line was reported earlier this month by a technician working on wiring inspections in that area of Endeavour's cargo bay. The freon line is part of a cooling system for some of the SRTM electronics. Although the bent line has not shown any evidence of a leak, engineers are evaluating the possibility of either bracing or replacing the line to prevent further damage that could result from vibrations experienced at launch. Launch options are being reviewed this week for Space Shuttles Endeavour and Discovery. No specific target launch dates will be announced until a review of the wiring inspections is conducted later in the week and managers are satisfied with the progress of repairs. STS-103: Wiring inspections and repairs continue aboard the orbiter Discovery. The payload bay doors are open following their precautionary closure this weekend due to the potential for high winds from Hurricane Dennis. The crew for mission STS-103 will be at KSC later this week for their scheduled Crew Equipment Interface Test (CEIT). STS-101: Orbiter Atlantis is currently being stored inside the Vehicle Assembly Building, awaiting the opening of an Orbiter Processing Facility bay. Preliminary wiring inspections are complete and a more thorough inspection will begin once Atlantis is moved to the Orbiter Processing Facility. Preparations are underway at this time to move Atlantis from the VAB transfer aisle to VAB high bay 2 on Wednesday. Bruce Buckingham. (1999). **Kennedy Space Center Space Shuttle Status Report** [Online]. Available E-mail: domo@news.ksc.nasa.gov/subscribe shuttle-status [1999, August 31].]

DURING AUGUST: The failure of a Boeing Inertial Upper Stage during the Apr. 9 Titan IVB launch of a missile warning satellite at Cape Canaveral was caused when a thermal wrap and tape applied to a harness and connector prevented proper disconnection of a plug linking the first and second stages of the IUS, according to a U.S. Air Force investigation. The second stage was therefore still connected to the first stage at one location in a hinged clamshell fashion when the should have separated cleanly. When the second stage attempted to deploy its extendible nozzle and fire, the nozzle was damaged and the stage was in an improper attitude. This resulted in the $250-million Defense Support Program missile warning satellite being placed in a useless elliptical orbit. The Air Force has also detailed the cause of a May 8 incident on Launch Complex 17A here when a USAF/Lockheed martin GPS spacecraft received $2.1 million in water damage from rain penetrating its clean room atop the pad. The Air Force determined that rain penetrated the clean room because of "the structural condition of the room where the satellite [was] kept and the lack of process to discover holes in the room." The mishap review board also determined that "a failure to follow procedures for fabricating and installing" a rain shield around the satellite also allowed the rainwater that penetrated the clean room to reach the satellite. ["Plug Wrap, Procedures Let to IUS Failure," **Aviation Week & Space Technology**, August 23, 1999, p 34.]

◆   The Hughes Space and Communications Co. has asked NASA, on behalf of international insurance underwriters, to launch the shuttle on a commercial space salvage mission to capture the wayward Hughes Orion 3 communications satellite so astronauts can attach a new upper stage rocket motor to the stranded satcom. The shuttle crew would then release Orion 3 so it could climb into geosynchronous transfer orbit, then stationary orbit, to complete its communications mission. Loral, which would have owned the spacecraft, received a $265-million insurance payoff this month on the failed Delta III mission that stranded Orion 3. Insurance underwriters now own the spacecraft. If reconfigured by shuttle astronauts, then boosted as intended, Orion 3 has the potential of earning $1 billion or more over more than 10 years of service life for whatever communications company the satellite is sold to after placed into proper orbit. If the shuttle rescue of Orion 3 is approved by NASA, it would return the shuttle program to the same type of daring commercial space salvage operations it performed in the mid-1980s and early 1990s. ["Space Shuttle Rescue Sought for Orion 3," **Aviation Week & Space Technology**, August 23, 1999, p 36.]

SEPTEMBER 2: Space Shuttle Status Report, Thursday, September 2, 1999. Shuttle managers today reviewed the progress of electrical wiring inspections and repairs on Endeavour and Discovery. Although the work is progressing well, evaluations of the findings thus far are continuing and managers have expanded the inspections based on those findings, including further inspections of areas below the floor of the payload bay. The time required to complete the work is still being assessed. Managers do not plan to discuss target launch dates for upcoming missions until more of the work has been completed, however it is anticipated that no mission could technically be ready for launch before mid-October at the earliest. Although damage to wiring has been found and repaired in each orbiter, the primary focus of the inspections and repairs is to put measures in place that ensure damage to wiring does not recur. Those measures include installing flexible plastic tubing over some wiring, smoothing and coating rough edges in the proximity of wiring, and installing various other protective shielding where needed. Also, the ground procedures and equipment used when preparing the orbiters for flight are being revised to reduce the potential for technicians to cause inadvertent damage, and plans are being formulated to ensure that electrical wiring is subject to a thorough inspection regularly as part of standard shuttle maintenance. "Our focus is to be absolutely certain that we do our very best to find, fix and prevent any recurrence of this problem," Space Shuttle Program Manager Ron Dittemore said. "We will not discuss potential launch dates for upcoming missions until we are satisfied that we have done everything we need to do to fly safely. That has and will always be our top priority." When completed, the technicians will have inspected roughly 100 miles of electrical wiring in each of the four Shuttle orbiters, comprehensively covering the vast majority of the main electrical wiring from nose to tail of the spacecraft. The areas designated for inspection have been identified by gauging their susceptibility to damage by the amount of work generally performed in those areas in the past; past modifications made to the orbiter in those areas; and the past record of wiring damage reports in those areas. Although numerous locations throughout the orbiters have been identified that require additional preventative measures, the number of places identified in each orbiter where wire has required repair includes: * Endeavour – 38; * Discovery – 26; * Atlantis -- full inspections will begin later this month; * Columbia -- other than initial inspections associated with the short experienced during STS-93, full wiring inspections will be performed when Columbia arrives at the Boeing North American shuttle factory in Palmdale, Ca., late this month. The repaired wires include areas such as connectors that require some additional insulation and other improvements. Less than half of those identified above are locations where wire repairs related to nicks and other inadvertent damage were required. The inspections have revealed no wiring problems related to age or wear factors. Payload engineers have determined that the bent freon line associated with the SRTM payload will be repaired with a brace and replacing the line will not be necessary. The bent line was reported earlier this month by a technician working in that area. The freon line is part of a cooling system for some of the SRTM electronics. Orbiter Atlantis is currently being stored inside the Vehicle Assembly Building, awaiting the opening of Orbiter Processing Facility bay 3. Wiring inspections will begin once Atlantis is moved to the OPF. Orbiter Columbia continues to undergo routine post flight-deservicing in Orbiter Processing Facility bay 3. Workers are preparing the orbiter for its upcoming Orbiter Maintenance Down Period (OMDP) in Palmdale, CA. Once at Palmdale, extensive wiring inspections will be conducted. Columbia is scheduled to be mounted atop NASA's modified Boeing 747 on Sept. 22. The ferry flight from KSC is scheduled to begin on Sept. 23 with an overnight stop at Luke Air Force Base in Arizona. Current plans have Columbia arriving in Palmdale on Sept. 24. Because the orbiter can not be flown through precipitation of any kind, ferry flight plans are contingent upon weather conditions in the flight path. Bruce Buckingham. (1999). **Kennedy Space Center Space Shuttle Status Report** [Online]. Available E-mail: domo@news.ksc.nasa.gov/subscribe shuttle-status [1999, September 2].]

◆ The third International Space Station (ISS) flight, STS-101, has been rescheduled for launch of Jan. 22, 2000. The 11-day mission is expected to return to KSC on Feb. 2, 2000. STS-101 will include two EVAs for external equipment transfer. [**KSC Countdown**, September 2, 1999.]

◆ Nude sunbathers have quietly basked in the sun at Playalinda Beach this summer without being ticketed for shedding their clothes. The citation-free summer is the first since 1995, when Brevard County commissioners passed an ordinance prohibiting thongs, T-backs and public nudity. Last year, Brevard sheriff's deputies issued tickets or arrested 16 nude sunbathers at Playalinda Beach, which is at Canaveral National Seashore, a federal property. In 1997, 60 people were arrested or ticketed. ["Nude Playalinda sunbathers go ticketless," **Florida Today**, September 3, 1999, p 1A.]

◆ When Congress returns after Labor Day, one of the first orders of business will be a spending bill that cuts $1 billion from the NASA budget, a move that could bring more layoffs. In its current form, the bill gives NASA $12.6 billion. That's enough to keep the costly International Space Station project going by cutting other missions, including a satellite Vice President Al Gore wants launched so it can beam pictures of Earth to the Internet. NASA supporters are hopeful that between now and Oct. 1, the start of the new fiscal year, money will be found to boost the civilian space agency's budget allocation. But if the cut sticks, thousands of NASA workers could lose their jobs, space agency Administrator Daniel Goldin has said. President Clinton had asked for $13.5 billion for NASA in fiscal year 2000. The space agency's budget for the current fiscal year is $13.6 billion. ["NASA backers hope House won't Ok cuts," **Florida Today**, September 3, 1999, p 1A & 2A.]

SEPTEMBER 4: In what would be one of the most significant national space policy changes in 50 years, the Clinton administration is moving to take away total control of U.S. launch ranges from the military. The goal: to run the ranges more like civilian airports. That would give U.S. industry the flexibility to carry out national security missions for the pentagon and science missions for NASA, while vying for a bigger share of the global market to launch commercial satellites. The U.S. Air Force for five decades has run the federal government's Eastern and Western launch ranges, which are based at Cape Canaveral Air Station and Vandenberg Air Force Base in California. Each of the ranges includes widespread networks of ground stations used to track rockets in flight and destroy them if they stray off course and threaten public safety in the United States or foreign countries. Long the exclusive domain of the Department of Defense and NASA, the ranges now host more commercial launches than military or civil government missions. That trend is expected to continue because of global demand for satellite television, telephone and data relay services, and the fact U.S. rocket companies will launch most Pentagon and NASA missions during the 21st century. In the near term, the Air Force would be ordered to "establish a more businesslike management organization to balance national security, civil and commercial interests," the report states. Within a couple of years, the White House would establish a new federal office – jointly managed by the Air Force, the Federal Aviation Administration and perhaps NASA – to run the ranges and protect the public from runaway or exploding rockets. ["Military's role in running ranges may be reduced," Florida Today, September 5, 1999, p 1A & 2A.]

◆　Thiokol Propulsion has signed a $1.7 billion contract extension to build the space shuttle's solid rocket boosters through 2005. "This says we're going to be building solid rocket motors for the shuttle for a long time," said Bob Crippen, president of Thiokol Propulsion, a division of Salt Lake City-based Cordant Technologies. Crippen, a former astronaut, said the company expects future extensions and NASA has said that it will fly the shuttle at least through 2012. The new contract is the sixth in a series of pacts awarded to Thiokol for the design, development, production and refurbishment of shuttle boosters. Under its provisions, Thiokol will use new and refurbished hardware to provide 70 flight motors – enough for 35 space shuttle flights – and three test boosters. Delivery of the first flight booster is scheduled for this month. Thiokol has delivered 146 flight rockets since 1988, when NASA reinstituted shuttle flights following the Challenger explosion, blamed on the failure of seals on the solid rocket boosters. The boosters were redesigned after the accident and have performed without serious problems. ["Thiokol signs contract extension for space shuttle rocket boosters," Florida Today, September 5, 1999, p 1E.]

SEPTEMBER 6: If proposed spaceship called VentureStar flies someday, Florida's aerospace leaders say they are more confident that ever it will do so from the state's launch pads. But they admit there's one nagging problem in the 15-state competition to be the launch site for the Lockheed Martin vehicle. It may never get off the ground. That's why, although they're focused on VentureStar right now, state officials say they will broaden their net to try to attract any future launch vehicles. They hope to catch whatever actually makes it into space. Concerns stem from Lockheed Martin's delays in building the X-33 vehicle that is meant to test key technologies for the VentureStar spaceship. The X-33 is a half-sized model of VentureStar and must fly successfully before Lockheed Martin decides whether to continue developing the larger, more complex spaceship. Set to begin this summer, the X-33's test flights were pushed back a year to July 2000 because more time is needed to build the advanced engines and lightweight fuel tanks that will be used on the vehicle. NASA – which has paid Lockheed Martin more than $900 million to use toward the X-33 – would be a prime customer because VentureStar could ferry supplies and crews to the International Space Station. Around the year 2012, VentureStar could replace the agency's space shuttles completely when they retire after 30 or more years of service. ["Officials sure state can lure VentureStar," <u>Florida Today</u>, September 7, 1999, p 1A & 2A.]

SEPTEMBER 7: NASA has pulled one of its most experienced spacewalkers from a space station construction mission for undisclosed reasons, officials said Tuesday. The astronaut, Mark Lee, is a veteran of four flights with more than 26 hours of spacewalks, including three excursions to improve the Hubble Space Telescope in 1997. NASA officials won't say why the 47-year-old Air Force colonel was replaced, or what duties to which he is currently assigned. "It's an internal astronaut office matter, and because of that I'm not a liberty to discuss the details," said Ed Campion, a spokesman at NASA's Johnson Space Center in Houston. Lee has been replaced by astronaut Bob Curbeam, who has flown once and was training to perform spacewalks on another station flight. NASA officials say the switch won't affect the mission. ["Veteran spacewalker bumped from mission," <u>Florida Today</u>, September 8, 1999, p 1A.]

SEPTEMBER 9: Space Shuttle Status Report, Thursday, September 9, 1999. NOTE: Although inspections and repairs of electrical wiring are continuing, Space Shuttle managers today announced tentative target launch dates for the next two Shuttle missions, STS-103 and STS-99. Managers established STS-103, the Hubble Space Telescope Servicing Mission 3A, as the next Shuttle flight with a launch targeted for no earlier than Oct. 28, 1999. A target launch date of no earlier than Nov. 19, 1999, was set for the launch of STS-99, the Space Radar Topography Mission. Managers have directed all those involved in flight preparations to work toward these dates, however the launch dates remain tentative pending the outcome of the current wiring work. "We've established these planning dates so that those involved in flight preparations can establish the proper order of priority for work on each mission," Space Shuttle Program Manager Ron Dittemore said. "However, we are continuing to review the progress of wiring inspections and repairs on Endeavour and Discovery and, as part of our continuing evaluations of those activities, we may need to revisit the subject and further adjust our target launch dates as those repairs progress. We will not fly any mission until we are satisfied that we have safely resolved the all wiring problems we have found." STS-103: Wiring inspections and repairs continue aboard Discovery. Most of the work in the forward crew and aft engine compartments is complete. Emphasis is now being concentrated on work in the midbody. Also, the Crew Equipment Interface Test (CEIT) was successfully conducted late last week. STS-99: Wiring inspections and repairs continue in the orbiter's midbody and aft engine compartment. Inspections and repairs in the forward crew compartment are complete. Also, fabrication of the SRTM freon line reinforcement clamp assembly has begun. The manufacturing of the clamp was done at KSC and will be ready this week for a fit check. The fixture is scheduled to be installed for flight in the payload bay next week. STS-101: Orbiter Atlantis is currently being stored inside the Vehicle Assembly Building awaiting the opening of an Orbiter Processing Facility bay. Preliminary wiring inspections are complete. A more thorough inspection will begin once Atlantis is moved to the Orbiter Processing Facility. Last Wednesday, Atlantis was moved from the VAB transfer aisle to VAB high bay 2. Preparation for Orbiter Maintenance Down Period (OMDP): Initial wiring inspections are complete on the orbiter Columbia in Orbiter Processing Facility bay 3. Work continues to prepare the vehicle for its ferry flight to Palmdale, CA, for its scheduled Orbiter Maintenance Down Period (OMDP). Columbia is scheduled to be towed to the Shuttle Landing Facility on Sept. 22 where it will be mated to the modified Boeing 747. The ferry flight from KSC is scheduled to begin on Sept. 23 with an overnight stop at Luke Air Force Base in Arizona. Columbia will depart Luke the following day, arriving in Palmdale on Sept. 24. Ferry flight plans are contingent upon weather conditions in the flight path and the vehicle could be diverted to other facilities with little notice. Bruce Buckingham. (1999). **Kennedy Space Center Space Shuttle Status Report** [Online]. Available E-mail: domo@news.ksc.nasa.gov/subscribe shuttle-status [1999, September 9].]

◆ Turning aside complaints from pro-space legislators, the House of representatives cut $1 billion from NASA's spending bill on Thursday. Now the measure moves to the Senate with the thorny question of giving more money to the civilian space agency at the expense of veterans' health care and low-income housing programs. NASA Administrator Daniel Goldin warned that a cut that deep would force thousands of layoffs and lead to closing at least two of the agency's nine space centers. Many pro-space legislators anticipate NASA will ultimately get the money back – possibly as part of a broad, 11th-hour spending deal between congressional leaders and the Clinton administration. Also, Thursday, the House International Relations Committee passed legislation to withhold U.S. assistance to support Russia's involvement in the International Space Station unless the White House certifies Russian companies are not transferring missile and weapons technology to Iran. ["House trims $1 billion off NASA budget," __Florida Today__, September 10, 1999, p 1A & 10A.]

SEPTEMBER 10: As Lockheed Martin prepares one Atlas rocket for launch Monday after a four-month hiatus, engine concerns have again sidelined two sister vehicles the company once said were cleared for flight. Officials say the second round of inspections of an Atlas rocket slated to launch a navy communications satellite later this month has raised new concerns about one of the two upper stages engines. The launch has been put on hold, and the suspect engine may have to be replaced. In addition, an Atlas set to launch NASA's Terra environmental research satellite had received approval for launch by Lockheed Martin but has been grounded again. The reason: NASA wants its upper stage engines reinspected. The space agency has not yet approved the method Lockheed Martin is using to clear the engines for flight. The one Atlas rocket that has been given a clean bill of health is awaiting liftoff from Cape Canaveral early Monday carrying a TV broadcasting satellite. The launch window will extend from 1:11 to 3:10 a.m. Air Force meteorologists are predicting an 80 percent chance of good weather. ["Atlas rocket launch set for Monday," **Florida Today,** September 11, 1999, p 2A.]

SEPTEMBER 11: The chief of Florida's 10 public universities will meet with Kennedy Space Center Director Roy Bridges on Tuesday (September 14, 1999) to talk about the state's growing interest in the commercial space industry. University Chancellor Adam Herbert wants to discuss the Florida Space Research Institute, a consortium of public and private universities, industry leaders and state officials the Legislature created this year. ["Universities vie for space research," <u>Florida Today</u>, September 12, 1999, p 1A.]

SEPTEMBER 13: Space Shuttle Status Report, Monday, September 13, 1999. Hurricane Floyd Preparation Status: Over the weekend, KSC workers got a head start on implementation of the center's standard hurricane preparedness plan. By about 9 p.m. tonight, all possible steps to protect Shuttle flight hardware, payloads, equipment and facilities will be accomplished at KSC. Should Hurricane Floyd continue on its predicted track, all KSC employees will be released from work at about 4:30 p.m. today to allow adequate time for personal hurricane preparation. Forecasters currently indicate that Central Florida's east coast could experience winds in excess of 50 knots by 4 p.m. Tuesday and in excess of 100 knots by 2 a.m. Wednesday. Space Shuttles Discovery, Endeavour and Columbia remain in Orbiter Processing Facility bays 1, 2 and 3 respectively. Orbiter Atlantis is being stored in Vehicle Assembly Building high bay 2. Today, workers closed the payload bay doors on all orbiters and are protecting the landing gear as part of standard hurricane protection efforts. Shuttle and payload test equipment will be raised above floor level to avert flood damage. The Rotating Service Structures at Launch Pads 39A and 39B will be rotated back to the Fixed Service Structures to maximize their protection from high winds. Eleven solid rocket booster segments are being prepared for transport to Tallahassee, FL, by railroad. If railroad traffic and local evacuation efforts prohibit transport efforts, the segments will be secured at KSC underneath their railcar covers. The Vehicle Assembly Building and launch pads can withstand winds of 125 mph. The Orbiter Processing Facility can withstand winds of 105 mph and other payload and flight hardware support facilities can endure winds of 110 mph. Besides protecting the orbiters and payloads within the facilities, the buildings themselves are being secured. This includes boarding windows, removing or tying down antennas, and sandbagging doors. The Kennedy Space Center elevation is approximately nine feet, so a concern for water intrusion exists in the event of a storm surge. The Orbiter Processing Facility is designed to withstand winds of 105 mph, the VAB, PHSF and launch pads 125 mph, and the other payload and flight hardware facilities, 110 mph. In the Space Station Processing Facility, the International Space Station flight hardware is already elevated in test stands but are also being covered. Test equipment is also being elevated and covered. The SRTM payload will ride out the storm in the Space Station Processing Facility high bay inside the payload canister with the doors closed. In the Payload Hazardous Servicing Facility, the Hubble Space Telescope flight elements are being bagged and the test equipment covered. Rideout crews totaling about 120 individuals will be stationed throughout KSC during the storm. At about 4 p.m. Tuesday, a decision will be made to keep the rideout crews on location at KSC or to have then evacuate the center as well. KSC is scheduled to be closed Tuesday and Wednesday. Bruce Buckingham. (1999). **Kennedy Space Center Space Shuttle Status Report** [Online]. Available E-mail: domo@news.ksc.nasa.gov/subscribe shuttle-status [1999, September 13].]

SEPTEMBER 14: Eighty workers volunteered to stay behind at the evacuated Kennedy Space Center on Tuesday to ride out Hurricane Floyd, which NASA feared could destroy the launch pads and shuttle hangars. Forecasters warned the storm could pass 35 miles to 50 miles offshore, bringing wind of 130 mph Wednesday morning. NASA evacuated its approximately 12, 500 workers Monday and closed down the space center. It also reduced its usual skeleton staff of 120. "Everybody else is gone. It's kind of eerie out here," NASA spokesman George Diller, one of the volunteers, said by telephone from a fortified building at the space center. "I kind of can feel the concern growing." NASA feared not only the wind, but the storm surge: The space center is barely nine feet above sea level. And the facility's space shuttle hangars and launch pads are designed to withstand wind of no more than 125 mph. All four shuttles, worth $2 billion apiece, are parked indoors. At the nearby Cape Canaveral Air Station, four rockets stand exposed to the storm on their launch pads. The next shuttle mission is scheduled for the end of October but probably will be delayed because of the interruption in work. ["Small group of volunteers stays at space center," **Florida Today**, September 15, 1999, p 2A.]

SEPTEMBER 15: A deep cut in spending that threatened to wreak havoc for NASA was eliminated Wednesday in the Senate's version of next year's federal budget. The House voted to cut $1 billion from the agency's existing $13.6 billion budget for space research and exploration. But the Senate Appropriations subcommittee for space, veterans programs and housing put the money back in Wednesday by raiding the human-services budget. Although the transfer will set off other budget problems in Congress, top senators on the committee were optimistic that NASA's budget, held to zero growth in recent years, will be spared a reduction when the next fiscal year starts in October. The full Senate will vote on the budget as early as next week. A conference committee will then work out differences with the House later this month. ["Senate restores $1 billion to NASA," **The Orlando Sentinel**, September 16, 1999, p A-6.]

◆ Kennedy Space Center, Cape Canaveral Air Station and Patrick Air Force Base reported only superficial damage from Hurricane Floyd on Wednesday. All three are expected to operate normally today. NASA's shuttle fleet was kept safe inside facilities at KSC, where Hurricane Floyd caused only minor damage to the center's grounds and some structures. Water was blown under the door of a hangar where three of NASA's $2 billion spaceships are being prepared for future missions. But it didn't reach the shuttles. "God has been good to us," NASA Administrator Dane Goldin told CNN early Wednesday. Officials had worried that the spaceships could have been damaged extensively if Floyd had made a direct hit in the region with winds of 140 mph. None of the shuttles were in facilities built to withstand winds greater than 125 mph. Goldin said "the space agency will have to look hard" at the precautions it should take for future hurricanes. "This rings a warning bell," he said. At Cape Canaveral Air Station, the four rockets that were left outside on their launch pad were not outwardly damaged by the storm. The rockets are protected by launch towers that can withstand winds up to 120 mph. "From the outside, there doesn't appear to be any damage," said Lynda Yezzi, a spokeswoman for the 45th Space Wing at Patrick Air Force Base. ["NASA, Air Force report superficial damage," Florida Today, September 16, 1999, p 1A.]

SEPTEMBER 17: Space Shuttle Status Report, Friday, September 17, 1999. NOTE: Kennedy Space Center employees returned to work Thursday from two days of administrative leave after Hurricane Floyd headed north along the east coast of Florida. The Space Shuttle orbiters and associated flight components were undamaged by the high winds and rain. Also, no damage was reported to any other flight hardware, the International Space Station elements, the SRTM payload, or the Hubble Space Telescope servicing mission components. The highest winds recorded were 91 mph from the NNW at 4:50 a.m. on Wednesday at a weather tower located between Shuttle Launch Pad 39A and Launch Complex 41. The maximum sustained winds were recorded at 66 mph. The highest amount of rain recorded at KSC was 2.82 inches as the eye of Hurricane Floyd passed 121 miles east of Cape Canaveral at 4 a.m. Wednesday. Minor damage at KSC included numerous signs, some trees, and two traffic lights blown over. Some VAB siding panels were blown off on the east and west sides of the building, however, no structural damage occurred. There was minor damage at pad 39B to the weather protection panels that enclose the pad's primary electrical system. Some water intrusion occurred underneath hangar doors at the Orbiter Processing Facility high bays, the doors of the adjacent main engine maintenance facility, and the north door of the VAB transfer aisle. STS-103: Yesterday afternoon, Shuttle managers reviewed the progress of ongoing wiring inspections and repairs to orbiters Endeavour and Discovery and adjusted the planning date for launch of the next Space Shuttle mission to no earlier than Nov. 19. Managers also decided to preserve the option to launch either STS-99, the Space Radar Topography Mission or STS-103, the Hubble Space Telescope Servicing Mission 3A, as the next flight. Although a few work days at the Kennedy Space Center were lost due to Hurricane Floyd, the primary rationale for adjusting the planning launch dates are the continuing inspections and repairs of electrical wiring. Technicians are still in the midst of conducting the work on Endeavour and Discovery, including inspections of wiring located under the liner of the payload bay that were added to the plans earlier this month. Managers plan to meet again next week to further review the progress of the work and possibly set new target launch dates. Managers also conducted a comprehensive review of the electrical wiring on other Shuttle components, including the solid rocket boosters, external tank and main engines, and found no need for further inspections of those components. STS-101: Orbiter Atlantis remains inside Vehicle Assembly Building high bay 2, awaiting the departure of Shuttle Columbia from OPF bay 3. Managers plan to move Atlantis to the OPF next week. While in the VAB, Atlantis has undergone midbody wire inspections. Further wire inspection and repair will be conducted in the OPF. Preparation for Orbiter Maintenance Down Period (OMDP): Orbiter Columbia's departure for Palmdale, CA has been delayed as a result of Hurricane Floyd's threat and KSC's mandatory evacuation. Final preparations for the cross-country ferry flight are under way and managers plan to move Columbia to the Mate/Demate Device at the Shuttle Landing Facility on Friday. Columbia will be mounted atop NASA's modified Boeing 747 on Sept. 24 and departure is currently slated for Sept. 25. An overnight stop at Luke Air Force Base is currently scheduled to accommodate refueling requirements for the Shuttle Carrier Aircraft. Current plans have Columbia arriving in Palmdale, CA on Sept. 26. Because the orbiter can not be flown through precipitation of any kind, ferry flight plans are contingent upon weather conditions in the flight path. Bruce Buckingham. (1999). **Kennedy Space Center Space Shuttle Status Report** [Online]. Available E-mail: domo@news.ksc.nasa.gov/subscribe shuttle-status [1999, September 17].]

SEPTEMBER 18: NASA's stalled International Space Station construction project soon will get a major jump-start with the long-delayed launch of a crucial Russian command and control module. Already more than a year behind schedule, Russia's Zvezda module, which will double as living quarters for early station crews, is scheduled for launch Nov. 12 from Baikonur Cosmodrome in Kazakstan. ["Crucial station piece set for Nov. launch," <u>Florida Today</u>, September 19, 1999, p 1A.]

SEPTEMBER 19: A NASA name tag coated with lunar dust that was worn by an astronaut who walked on the moon sold at auction for $310,500. The 6-by-12-inch cloth keepsake was cut from an insulated jacket worn by an astronaut, the late James Irwin, during the 1971 flight of Apollo 15. The lunar dust – which created a dark gray tint around the raised edges of the tag – became embedded into the tag during three separate moon walks Irwin took. The item, which was sold Saturday by Christie's auction house on behalf of Irwin's estate, sold for three times its presale estimate and was the highest-selling lot in the sale of nearly 300 space-related items. ["NASA tag with moon dust sells for $310,500," <u>Florida Today</u>, September 20, 1999, p 2A.]

SEPTEMBER 20: NASA and Boeing are celebrating the opening of a new NASA Technical Records Center in a ribbon-cutting ceremony on Tuesday, September 21, at 10:30 a.m. The new facility, completed in June, solved several space issues for NASA and Boeing through an innovative agreement and excellent teamwork. Boeing gained the space they needed to support their new Delta IV program and NASA gained a new records storage site. Center Director Roy Bridges and other officials will be attending the ceremony. ["Ribbon cutting ceremony for NASA Technical Records Center," **NASA News Release #73-99**, September 20, 1999.]

◆ NASA on Monday launched an investigation into the way maintenance work is done on its space shuttle fleet, a move prompted by the discovery of damaged electrical wiring on all four vehicles. The problems were detected during inspections that were ordered after an electrical short knocked out two main engine computers during a July 23 shuttle launch. NASA space flight chief Joe Rothenberg tapped Henry McDonald, director of NASA's Ames Research Center near Mountain View, Calif., to head the probe. His job: Review all shuttle maintenance practices at Kennedy Space Center as well as work done when the vehicles are periodically shipped to California for modifications. McDonald will form a team that includes top maintenance experts from NASA, the military and the commercial airline and aerospace industries. Preliminary findings are due in October. ["NASA team to probe shuttle work," Florida Today, September 21, 1999, p 1A]

◆ Cautious but not alarmed, NASA and Air Force officials made some preparations Monday for the heavy rains expected today from Tropical Storm Harvey. At Kennedy Space Center, officials expected no threat to the four shuttles, which remain safely tucked in their hangars. Minor work was being done to other buildings. "We're concerned a little bit about the heavy rains, so we're sandbagging some doors in areas that are prone to (flooding), but overall, we don't expect anything serious," said George Diller, a KSC spokesman. As a precautionary measure, NASA did put plastic covers on some pieces of the International Space Station that are undergoing work inside a KSC facility. Four rockets are on their launch pads at Cape Canaveral Air Station, but all will be protected inside structures that can withstand winds up to 120 mph. Harvey's top wind gusts should be no stronger than 80 mph. Harvey did scuttle the planned launch early Wednesday of an Atlas rocket loaded with a TV satellite. Lockheed Martin officials will now try to launch the vehicle between 1:07 and 3:06 a.m. Thursday, weather permitting. ["NASA, Air Force prepare for Harvey," Florida Today, September 21, 1999, p 7A.]

SEPTEMBER 21: From outer space to cyberspace, veteran astronaut Sally Ride on Tuesday was appointed president of Space.com, the Internet site devoted to covering the news and science of space. ["Space.com chooses astronaut as president," **The Orlando Sentinel**, September 22, 1999, p A-7.]

SEPTEMBER 22: Space Shuttle Status Report, Wednesday, September 22, 1999. STS-103: Inspections of the wiring in Discovery's forward and aft compartments are complete. Inspection and repair of the wiring in Discovery's midbody, above the payload bay liner, are very near completion. About 50 percent of wiring inspections and repairs below the payload bay liner are complete. NASA's wiring inspection and repair efforts have focused on high traffic areas within the orbiter and in areas where work had recently been performed. This week workers replaced an oxidizer valve on Discovery's orbiter maneuvering system and wire protection work continues. Shuttle managers will meet Thursday afternoon to discuss the progress of the ongoing wiring maintenance effort, and to discuss mission priority and target launch dates for upcoming flights. STS-99: Orbiter Endeavour has undergone inspections of the forward, midbody and aft compartments and repair efforts above the payload bay liner are near complete. Wiring protection efforts are ongoing and inspections beneath the orbiter's payload bay liner will begin as access is established. Last weekend, workers replaced Endeavour's oxygen/nitrogen panel and today leak checks are under way. In the Space Station Processing Facility, the freon line reinforcement clamp will be installed on the SRTM payload later this week. STS-101: Orbiter Atlantis remains inside Vehicle Assembly Building high bay 2, awaiting the departure of Shuttle Columbia from OPF bay 3. Managers plan to move Atlantis to the OPF on Friday. While in the VAB, Atlantis has undergone midbody wire inspections and wireless video system installation is ongoing. Further wire inspection and repair will be conducted in the OPF. Preparation for Orbiter Maintenance Down Period (OMDP): Orbiter Columbia's departure for Palmdale, CA is currently scheduled to occur at about 7 a.m. Friday. Final preparations for the cross-country ferry flight are under way and managers plan to move Columbia to the Mate/Demate Device at the Shuttle Landing Facility early tomorrow morning. Columbia will then be mounted atop NASA's modified Boeing 747 for departure the next day. The planned overnight stop at Luke Air Force Base has been canceled. Following the crash of an F-16 at Luke on Monday and an expected one-day slip in Columbia's arrival in Arizona, Luke officials expressed Shuttle support concerns. The current ferry flight plan has Fort Worth, TX slated as its only refueling stop and indicates a Palmdale, CA arrival before sundown on Friday, Sept. 24. Because the orbiter can not be flown through precipitation of any kind, ferry flight plans are contingent upon weather conditions in the flight path. Bruce Buckingham. (1999). **Kennedy Space Center Space Shuttle Status Report** [Online]. Available E-mail: domo@news.ksc.nasa.gov/subscribe shuttle-status [1999, September 22].]

SEPTEMBER 23: NASA has yet to calculate the cost of the safety stand down that has idled its space shuttles, but wiring inspections and maintenance alone have cost about $350,000, a space agency official told a House subcommittee Thursday. Additionally, United Space Alliance, the private aerospace consortium maintaining the shuttles, will not receive an estimated $2.5 million payment it would have otherwise earned for the on-time delivery of an orbiter for the next scheduled shuttle launch. Former astronaut Michael McCulley, vice president and deputy program manager for the alliance, told members of the House Space and Aeronautics subcommittee his company accepts full responsibility for wiring damage found in Columbia, Discovery, Endeavour and Atlantis. He said the company is doing so even though the problems may stem from maintenance performed on the orbiters before USA took over operational responsibility for the craft. Thursday's hearing was the first official accounting lawmakers received since the decision was made earlier this month to ground the shuttle fleet pending a thorough inspection for damaged wiring. ["Shuttle wiring inspections, maintenance cost $350,000," <u>Florida Today</u>, September 24, 1999, p 4A.]

SEPTEMBER 24: Space Shuttle Status Report, Friday, September 24, 1999. STS-103: Based on the progress of wiring inspections and repairs on Discovery, Shuttle managers today decided to place priority on launching mission STS-103, the Hubble Space Telescope Servicing Mission 3A, as the next Shuttle flight. Discovery is now planned for launch no earlier than Nov. 19, although a target launch date will continue to be assessed as the inspections and repairs continue. Because the area under the liner of Discovery's payload bay has been more easily accessible, the planned wiring work has progressed quicker on Discovery than it has on Endeavour. The decision to plan STS-103 as the next Shuttle flight was based on that progress. STS-99: Managers also continue working toward a possible launch of Endeavour this year on mission STS-99, the Space Radar Topography Mission. However, according to the current status of wiring work on Endeavour, it is projected that it could be ready for a launch of STS-99 no earlier than December. Managers plan to continue to assess the progress of the wiring inspections and repairs on both orbiters weekly and to adjust target launch dates once the time required to complete the work is better understood. STS-101: Atlantis was moved from the VAB to Orbiter Processing Facility bay 2 this morning. In the OPF, the vehicle will undergo wire inspections similar to those of Discovery and Endeavour. Preparation for Orbiter Maintenance Down Period (OMDP): Columbia's ferry flight to Palmdale, CA, began today for its regularly scheduled Orbiter Maintenance Down Period (OMDP). The orbiter, atop the modified Boeing 747 Shuttle Carrier Aircraft (SCA), departed KSC's Shuttle Landing Facility at 12:15 p.m. en route to Whiteman Air Force Base, MO, which is about an hour southeast of Kansas City, MO. Columbia landed at Whiteman at about 3:30 p.m. Eastern Time and will remain there overnight. Tomorrow the SCA will depart for Palmdale, CA., and arrive at about 1 p.m. Eastern Time. Ferry flight plans are contingent upon weather conditions in the flight path, and the vehicle could be diverted to other facilities with little notice. Bruce Buckingham. (1999). **Kennedy Space Center Space Shuttle Status Report** [Online]. Available E-mail: domo@news.ksc.nasa.gov/subscribe shuttle-status [1999, September 24].]

◆ When Columbia made the first shuttle flight in 1981, the future looked promising for the U.S. space program – and for NASA subcontractor USBI Co. Eighteen years later, as the sounds of Columbia's engines faded during its July flight into orbit, so did USBI's involvement in the program. On Thursday, the USBI legacy ends as NASA prime contractor United Space Alliance takes over USBI's role and absorbs 830 of its employees. That will boost United Space Alliance into a tie with Harris Corp. as Brevard County's largest private employer, with 6,000 local employees. ["Rocket contractor becomes just a memory next week," <u>Florida Today</u>, September 25, 1999, p 1A & 2A.]

◆ Two Kennedy Space Center employees are among 22 people honored this year by the Carnegie Hero Fund Commission for acts of heroism. Jerry Bowman, 33, and Daryl Elder, 37, braved alligator-infested waters to save 50-year-old James Swanson from his partially submerged car in Titusville on April 16, 1997, the commission said.

Bowman and Elder easily measured up to the commission's rigorous standards on the drizzly day they plucked Swanson from his car, which had veered into a deep canal that runs parallel to the NASA Causeway. Elder had been about waist-deep in the water when Bowman, standing on the shore, saw an alligator gliding toward Swanson's car. As the seconds ticked away, Elder jumped on the roof of the car and used a hammer to break the rear window. Bowman waded through the water and unlocked the car's door. The pair eventually pried Swanson from the car and dragged him to safety. Bowman and Elder each will receive a $3,000 check and a bronze medal for their bravery. ["Group honors two KSC workers for heroism," **Florida Today**, September 25, 1999, p 1A.]

◆ Shuttle Columbia is on its way to a California plant today to lose weight and get a $75 million makeover. By the time it's back next summer, NASA's oldest spaceship will have new equipment in the cockpit, a fresh paint job and a thorough inspection on about 100 miles of internal wiring. Attached to the back of a Boeing 747 jet, Columbia was flown Friday from Kennedy Space Center to Whiteman Air Force Base in Missouri. Its final destination is the Boeing Corp. plant in Palmdale, Calif. ["Columbia heads west for an overhaul," **Florida Today**, September 25, 1999, p 1B.]

SEPTEMBER 28: The 1998 class of astronaut candidates (referred to as ASCANs) is at KSC this week for training. It's the first visit for some of them. Among the week's activities are fire training and a flight awareness program, plus touring the Apollo/Saturn V Visitors Facility, the OPF, SSME Processing Facility, VAB, SSPF, launch pad, SLF and the crew head-quarters. ["Astronaut candidates touring KSC," **KSC Countdown**, September 28, 1999.]

◆ An uncomfortably close call with Hurricane Floyd is prompting NASA to look at the way it safe-guards the nation's $8 billion space shuttle fleet from such devastating storms. While the agency was moving to better protect its ships prior to Floyd, officials say little can be done to withstand the fury of the most powerful hurricanes. The vulnerability of the fleet – as well as shuttle hangars, launch pads and payload processing buildings – was pointed out as Floyd battered the Bahamas and then made a beeline toward Cape Canaveral. Even larger than 1992's deadly Hurricane Andrew, Floyd packed winds topping 150 mph as NASA hurried to secure its four orbiters in shuttle hangars and the 52-story Vehicle Assembly Building at KSC. Problem was, the buildings and launch pads only are capable of withstanding winds between 105 mph and 125 mph. Had Floyd not turned north, the storm could have done serious damage to the shuttles and the nation's space program. As it turned out, the eye of the storm passed 121 miles east of Cape Canaveral at 4 a.m. Sept. 15. Top sustained winds at KSC were 66 mph with gusts up to 91 mph. Damage was minor. Signs, trees and two traffic lights were blown down. Some weather-protection material came off a shuttle launch pad, and about 30 sheets of metal siding blew off the VAB. In the wake of the storm, though, Goldin ordered a review aimed at determining whether anything can be done to shore up shuttle hangars and launch facilities. The shuttle facilities review team is expected to make recommendations to NASA space flight chief Joe Rothenberg before the end of the year. ["NASA studies storm safety at space center," <u>Florida Today</u>, September 29, 1999, p 1A & 2A.]

SEPTEMBER 29: Marcelite Harris is used to taking the lead. After joining the Air Force in 1965, Harris worked her way up the ranks to the highest echelons of the Pentagon and became the first African-American woman major general. Her last job: overseeing 125,000 people and a $260 billion inventory of some of the country's most advanced weapons. Today, she is the first Florida site director for United Space Alliance. The private company is taking over operations of NASA's shuttle fleet at Kennedy Space Center. Part of Harris' job is to make USA more visible in the community, increasing its involvement in local affairs and possibly expanding into new areas of the aerospace industry. ["Space alliance names director," <u>Florida Today</u>, September 30, 1999, p 14C.]

SEPTEMBER 30: Space Shuttle Status Report, Thursday, September 30, 1999. STS-103: Inspections of the wiring onboard orbiter Discovery are about 95 percent complete, and Shuttle managers expect all remaining wire inspections to conclude next week. During retests of Discovery's recently replaced right-hand orbiter maneuvering system (OMS) engine pod, technicians noted an oxidizer isolation valve that did not cycle properly. Subsequent valve inspections revealed minor corrosion of a manifold No. 5 oxidizer valve, and managers have decided to replace that valve before the STS-103 flight. As a precaution, technicians will conduct thorough inspections of all manifold No. 5 oxidizer and fuel valves located in both OMS pods and the orbiter's forward reaction control system. Inspection of these six valves will be performed in parallel with ongoing orbiter processing and is not expected to further impact the schedule. In Discovery's payload bay, workers noted a worn latching mechanism on the housing for the foot restraint that space-walking astronauts install on the Shuttle's robot arm. Workers will replace the latching mechanism next week. Shuttle managers will meet again next Thursday to discuss the progress of the wiring maintenance effort, to receive an updated status on the oxidizer valve and latching mechanism work, and to assess Discovery's launch date. Space Shuttle Discovery will not launch before Nov. 19. STS-99: Inspections of the wiring onboard orbiter Endeavour are about 81 percent complete. Wiring protection efforts continue throughout the orbiter. Inspections beneath the orbiter's payload bay liner are in work as technicians gain access to the areas to be examined. The retesting of Endeavour's recently replaced oxygen/nitrogen panel continues. In the Space Station Processing Facility, the freon line reinforcement clamp has been installed on the SRTM payload. Shuttle managers continue to assess the option of launching Endeavour in December of 1999 and will discuss the STS-99 launch date next Thursday. STS-101: Orbiter Atlantis moved to Orbiter Processing Facility bay 3 Friday morning. This week, workers completed payload bay door functional tests. Preparations to install the right-hand OMS pod on the orbiter are in work and installation is slated for Monday. Wireless video modifications continue and wiring inspections are expected to begin in the orbiter's midbody tomorrow. Preparation for Orbiter Maintenance Down Period (OMDP): Orbiter Columbia arrived in Palmdale, CA, Sept. 25 just after 1 p.m. Eastern Daylight Time. Columbia's ferry flight atop NASA's Shuttle Carrier Aircraft took only one day with a single refueling stop at Whiteman Air Force Base, MO, on Friday afternoon. The orbiter stayed overnight at Whiteman and departed for Palmdale, CA, at about 9 a.m. Bruce Buckingham. (1999). **Kennedy Space Center Space Shuttle Status Report** [Online]. Available E-mail: domo@news.ksc.nasa.gov/subscribe shuttle-status [1999, September 30].]

◆ The $125 million spacecraft that was destroyed on a mission to Mars last week was probably doomed by NASA scientists' embarrassing failure to convert English units of measurement to metric ones, the space agency said Thursday. The Mars Climate Orbiter flew too close to Mars and is thought to have broken apart or burned up in the atmosphere. NASA said the English-vs.-metric mixup apparently caused the navigation error. ["A math error of galactic proportions," <u>The Orlando Sentinel</u>, October 1, 1999, p A-1 & A-14.]

OCTOBER

OCTOBER 1: The latest delay for NASA's International Space Station was announced Friday when agency officials said the Russian-built living quarters – on which all further construction hinges – won't be launched before Christmas and could be delayed until mid-January. The decision was made after meetings this week in Moscow with the Russian Space Agency, which had planned to launch the Service Module on Nov. 12. Now the segment won't fly until sometime between Dec. 26 and Jan. 16. ["Yet another space station delay," Florida Today, October 2, 1999, p 1A.]

OCTOBER 3: When a top NASA safety officer told a congressional panel recently that the space shuttle fleet could operate safely with as few as two flights a year, the eyebrows of skeptical lawmakers ascended in disbelief. Just two years earlier, another agency official told the same panel the magic number was five. "Any significant interruption in International Space Station assembly would drive the shuttle well below the five to six (per) year minimum rate recommended to maintain a safe, proficient team," said Wilbur Trafton, who was then NASA's associate administrator for space flight. With four orbiters idled for repairs to damaged wiring and still waiting for Russia to launch its key Zvezda service module component to the station, NASA has a different take on the flight rate. "We did a separate assessment and looked at, from a safety point of view, how few flights we could fly safely," said Frederick Gregory, a former shuttle commander who is now associate administrator for safety and mission assurance. "That number was about two." NASA has launched just two shuttle missions this year. That's the fewest since 1988, when two missions got off the ground as NASA returned the fleet to service following the Challenger explosion in 1986. With wiring inspections nearly complete on shuttle Discovery, agency officials are expected to make a decision soon to lift the launch moratorium. When that happens, Discovery will be used to fly a servicing mission to the Hubble Space Telescope in late November or early December, said Michael McCulley, vice president and deputy program manager for United Space Alliance, the private contractor responsible for shuttle ground operations. "Two a year is certainly safe." Said McCulley. Even if NASA could launch only one shuttle flight a year without an increased risk of human or mechanical error, no one would contemplate such a schedule because it would be a political, economic and public relations disaster for the space agency. The older flight rate study, directed by astronaut John Blaha, concluded eight shuttle flights a year was an optimum number, but added the rate could safely drop to six at reduced efficiency. The new study, requested by Gregory last year, concluded two was enough to keep the KSC shuttle workforce in a state of trained readiness. The study was directed by Michael Bloomfield, chairman of NASA's space flight safety panel. ["NASA: Two shuttle flights a year needed," Florida Today, October 4, 1999, p 1A & 2A.]

OCTOBER 4: Repairing the high-tech Space Mirror that memorializes America's fallen astronauts will require some high-tech investigation. Specifically, a linear accelerator, a device that can take X-rays through 16 inches of steel will be needed.

119

The 300-ton, $6.2 million Space Mirror at the Kennedy Space Center Visitor Complex has been shut down since January, when Astronaut Memorial Foundation officials heard a grinding noise as the mirror turned. Officials think the lubrication of a $240,000 two-ton ring on which the mirror rotates is the trouble but have been unable to figure out exactly where the problem is. Efforts to resolve the problem using equipment provided by NASA and its contractors have proved unsuccessful. The linear accelerator, to be brought here from California, is expected to provide a clear picture within 15 minutes. The X-rays will be shot while the Visitor Complex is closed. The procedure is expected to take about 2 ½ hours and cost $20,000 to $30,000. ["X-rays to seek mirror's flaw," <u>Florida Today</u>, October 5, 1999, p 1B & 2B.]

◆ Kennedy Space Center, the starting point for many missions to the far reaches of space, has crossed a less distant commercial frontier. A NASA-patented supersonic cleaning system technology developed at KSC has been transferred to a Dutch firm, marking the first time in KSC history that a U.S. patent owned by NASA has been licensed to a foreign company. ["NASA licenses first KSC technology to foreign company," **NASA News Release**, October 4, 1999.]

◆ A NASA subcontractor and its president pleaded guilty last week in federal court to overcharging the space program more than $16,000 for work on storage pallets. MET-CON Inc. of Cocoa and its president, Billy Sheffield, each pleaded guilty to one count of submitting a false and fraudulent claim to NASA. They will be sentenced in January. ["Subcontractor, its president plead guilty to defrauding NASA," <u>Florida Today</u>, October 5, 1999.]

OCTOBER 7: Space Shuttle Status Report, Thursday, October 7, 1999. With wiring inspections and repairs of Discovery and Endeavour nearing completion and similar work beginning on Atlantis, Shuttle program managers today set new planning target launch dates for the next three Space Shuttle missions. Based on an assessment of the work remaining on Discovery and Endeavour and the inspections which have begun on Atlantis, managers set the following as target launch dates for upcoming flights:

Target Launch Dates	*Mission/Shuttle*	*Payload*
Dec. 2, 1999	STS-103/Discovery	Hubble Space Telescope Servicing-3A
January 13, 2000	STS-99/Endeavour	Shuttle Radar Topography Mission
February 10, 2000	STS-101/Atlantis	ISS Logistics/Assembly Flight 2A.2
(no earlier than)		

"Our number one priority for the Space Shuttle is to fly safely, and that is why we delayed our launch preparations and have performed comprehensive wiring inspections and repairs," Space Shuttle Program Manager Ron Dittemore said. "As a result of our inspections, we've made significant changes in how we protect electrical wiring. We believe those changes, along with changes to the work platforms and procedures we use in the Shuttle's payload bay, will prevent similar wire damage from recurring," Dittemore added. STS-103: Inspections of the wiring on board Discovery should be complete by the end of the week as technicians wrap up final reviews of the wiring in the orbiter's crew module. In addition, engineers are completing the replacement of the manifold # 5 oxidizer isolation valve in Discovery's right hand orbital maneuvering system pod prior to its retesting to insure that there is no leakage in the system. The old valve was removed earlier this week after leakage was found within the pod. Rollover of Discovery from the Orbiter Processing Facility to the Vehicle Assembly Building is scheduled for no earlier than the end of October with its rollout to Launch Pad 39-B planned for no earlier than the first week in November. A launch on December 2 would occur at around 4:32 a.m. EST at the start of a 42-minute window. Landing would occur just after 2 a.m. EST on Dec. 12. A firm launch date will be set at NASA's Flight Readiness Review around the second week of November. STS-99: Inspections of the wiring on board Endeavour are about 90 per cent complete with no significant issues having been identified by engineers. The SRTM payload is expected to be reinstalled in Endeavour around October 22 with its rollover to the Vehicle Assembly Building planned for no earlier than the end of November. Rollout to Launch Pad 39-A is scheduled for around early December. Bruce Buckingham. (1999). **Kennedy Space Center Space Shuttle Status Report** [Online]. Available E-mail: domo@news.ksc.nasa.gov/subscribe shuttle-status [1999, October 7].]

◆ Lockheed Martin plans to blow up a Cape Canaveral launch complex, clearing the way for a new pad. Hundreds, perhaps thousands, of space workers will watch as the company carries out a launch pad demolition derby that will double as a fundraising party for charity. A live countdown will blare from loudspeakers. TV news choppers will be in the air. CNN, ABC, and Inside Edition all are expected to be on hand for what promises to be a quirky national media event. Lockheed Martin is preparing to bring on the most powerful version ever of its venerable Atlas rocket – the Atlas 5 – which will make its maiden voyage from Cape Canaveral's Launch Complex 41 in late 2001. Now 34 years old, the complex has been used to launch top-secret Pentagon satellites and famous NASA spacecraft, including the Viking Mars Landers and Voyager interplanetary probes, aboard Titan rockets. But the new Atlases won't need the old complex, so in a bid to save time and money its launch tower and rolling service structure will be toppled by 180 pounds of explosives. In less than a minute, 7 million pounds of steel will collapse in a mushrooming cloud of sand and dust. When the demolition is done, workers will spend a month hauling the steel debris to a recycling center. ["Historic launch facility's life to end with a bang," Florida Today, October 8, 1999, p 1A & 2A.]

◆ An Air Force Delta 2 rocket lifted off from Cape Canaveral Air Station carrying a military navigation satellite. Mounted atop a 12-story Delta 2 rocket, the $42 million Navstar Global Positioning System spacecraft blasted off at 8:51 a.m. The launch ended a string of failures, workplace accidents and weather delays. The mission marked the first successful Air Force launch from Cape Canaveral since May 1998. ["Air Force launch finally a success," Florida Today, October 8, 1999, p 2A.]

◆ NASA closed the book on a budget drama Thursday when Congress and the White House negotiated a deal to save the space agency from cuts this year. A House-Senate conference committee voted to give NASA $13.7 billion for the budget year that began Oct. 1 – a $100 million increase over last year's budget. The extra $100 million for NASA will go toward science programs and space shuttle safety upgrades. ["Deal saves NASA from budget cuts," <u>The Orlando Sentinel</u>, October 8, 1999, p A-3.]

OCTOBER 9: On Oct. 16 and 17, the KSC Visitor Complex will host a "Salute to Scouts" celebration. This year's activities will focus on both the history of space travel and its future. Called "Honor the Past, Imagine the Future," the event is open to Girl Scouts from all around Florida. By participating, girls are working toward their Aerospace Exploration Merit Badges. Scouts can complete a Mars Millennium Scavenger Hunt and earn a limited edition Kennedy Space Center patch and prize. The hunt asks girls to search for facts scattered throughout the many exhibits at the Kennedy Space Center Visitor Complex. ["KSC plans special weekend of activities for Girl Scouts," <u>Florida Today</u>, October 10, 1999.]

OCTOBER 14: Space Shuttle Status Report, Thursday, October 14, 1999.

Target Launch Dates	*Mission/Shuttle*	*Payload*
Dec. 2, 1999	STS-103/Discovery	Hubble Space Telescope Servicing-3A
January 13, 2000	STS-99/Endeavour	Shuttle Radar Topography Mission
February 10, 2000 (no earlier than)	STS-101/Atlantis	ISS Logistics/Assembly Flight 2A.2

STS-103: Less than 10 electrical panels remain to be inspected on Discovery's aft flight deck to conclude the orbiter's overall wire inspection effort. This work should take about 2 days and is expected to conclude early next week. Installation of wiring protection continues in other areas, and functional tests are under way. Replacement of the two leaky isolation valves on the right-hand orbital maneuvering system is complete. Loading of manifold No. 5 occurs this weekend, and thruster leak checks will follow. After the successful testing of Discovery's repaired wiring, Shuttle managers plan to transfer the orbiter from Orbiter Processing Facility bay 1 to the Vehicle Assembly Building on Oct. 28. With an on-time move to the VAB, Discovery could arrive at the launch pad by Nov. 3. A launch on December 2 would occur at around 4:32 a.m. EST at the start of a 42-minute window. Landing would occur just after 2 a.m. EST on Dec. 12. A firm launch date will be set at NASA's Flight Readiness Review on Nov. 19. STS-99: Endeavour's gaseous nitrogen servicing is complete. Installation of the orbiter's waste control system concluded yesterday. Wiring inspections on Endeavour are more than 90 percent complete, and wire protection and repair efforts continue. The SRTM payload is expected to be reinstalled in the payload bay around Oct. 22 with orbiter rollover to the Vehicle Assembly Building planned for no earlier than late November. Rollout to the launch pad is planned for early December. STS-101: Wiring inspections on Atlantis are in work. Workers plan to reinstall the right-hand orbital maneuvering system (OMS) pod on Saturday, following repair of an attach point this week. Inspections on the left-hand OMS pod are also scheduled. Ammonia boiler system work continues. Target launch dates for other Shuttle missions in 2000 will be reviewed by both Shuttle and International Space Station Program managers and discussed at an Oct. 21 meeting before being manifested for future vehicle processing. Bruce Buckingham. (1999). **Kennedy Space Center Space Shuttle Status Report** [Online]. Available E-mail: domo@news.ksc.nasa.gov/subscribe shuttle-status [1999, October 14].]

◆ In what proved to be a quirky media spectacle, hordes of reporters descended on the Cape by land and air to watch Lockheed Martin's destruction of Launch Complex 41 to a 7-million-pound mountain of twisted steel. Prior to the countdown, launch manager Adrian Laffitte addressed about 200 people, offering a bittersweet eulogy of sorts. In its heyday, Laffitte said, the complex was used to launch 27 Titan rockets that carried payloads including NASA's famous Viking Mars Landers and Voyager interplanetary probes. Lockheed Martin needed to raze the old launch tower and service station as part of a $250 million renovation effort necessary to launch the newest version of its Atlas rocket, the Atlas 5. ["Hundreds cheer last blast at pad," **Florida Today**, October 15, 1999, p 1B & 2B.]

OCTOBER 15: As a result of the threat from Hurricane Irene to the Kennedy Space Center and the Cape Canaveral vicinity, NASA management decided late today to cancel all Space Shuttle and payload work scheduled at KSC on Saturday. Though Irene will likely be a tropical storm by the time it makes closest approach to KSC, personnel scheduled to work on Saturday are being asked not to report during this period. There is a concern for a sustained wind of 40 knots with higher gusts and the probability of heavy rain that could flood some KSC roadways. Approximately 250 personnel are affected. ["Hurricane Irene postpones Saturday work at KSC," **NASA News Release**, October 15, 1999.]

OCTOBER 18: Space Shuttle Status Report, Monday, October 18, 1999. Hurricane Irene Status Report (updated): NASA conducted a preliminary damage assessment of Kennedy Space Center late Saturday after the passage of Hurricane Irene. The center of the storm passed approximately 35 miles east of Cape Canaveral at 2:30 p.m. on Saturday afternoon. The highest sustained wind velocity measured during the storm was 69 mph (60 knots) with a peak gust to 83 mph (72 knots) recorded at the wind tower near Complex 41. However, a sampling of other wind towers around the Complex 39 area showed the average sustained wind to be approximately 65 mph (56 knots). Total rainfall for the storm at KSC was 6.48 inches. Most of the damage sustained was similar in nature to that from Hurricane Floyd. There is damage to trailers, modular buildings, storage sheds and other structures of light construction. Trees and various signs are blown over with traffic lights malfunctioning, a few blown off their pedestals. Some additional siding panels were lost from the Vehicle Assembly Building. There were no problems within the building. There is light roof damage at the Launch Control Center and there was some minor leakage. This is also the case with various other buildings around KSC. There was some water intrusion under the OPF hangar doors from blowing rain but no roof leakage around the orbiters. There was some water intrusion in the payload changeout room at Pad 39-B but no damage to either Pad A or Pad B. There was no damage or leakage at the Space Station Processing Facility and work resumed there Saturday night. There was some minor roof leakage at the Payload Hazardous Servicing Facility where the HST 3A payload is located. The flight hardware was covered and elevated for the storm and not at risk. At the Visitor Complex, no tours were run Saturday and the facility closed to visitors at Noon. Some siding panels were lost from the IMAX Theater during the storm and there was some minor leakage in various facilities. As a precaution, NASA management directed that no scheduled work be performed at KSC from 6 a.m. to 6 p.m. on Saturday. As there is minimal processing activity scheduled on weekends, only about 250 people were affected. Bruce Buckingham. (1999). **Kennedy Space Center Space Shuttle Status Report** [Online]. Available E-mail: domo@news.ksc.nasa.gov/subscribe shuttle-status [1999, October 18].]

OCTOBER 21: Before committing large sums of money to upgrading NASA's space shuttle fleet, Congress should decide what the orbiters will do after the International Space Station is built, a space systems expert told a congressional panel Thursday. For the next four years, NASA's shuttles will be almost exclusively devoted to ferrying people, parts and supplies to the growing space station. But beyond that, there is no clear mission for the craft, said Stephen Book, an Aerospace Corp. engineer who participated in a recent National Research Council analysis of shuttle upgrades. "There ought to be a national policy on the shuttle before deciding on the upgrades," Book told members of the House space and aeronautics subcommittee. NASA has committed to keeping the shuttle fleet flying for at least the next decade. To meet that goal and continue flying safely, the agency is considering a number of enhancements and improvements that will make the shuttles more reliable and efficient. Although the shuttles and their main engine stacks look much as they did when Columbia first lifted off 18 years ago, the fleet has been in a constant state of improvement and repair, said William Readdy, NASA's deputy associate administrator for spaceflight. Readdy said NASA is reviewing its shuttle upgrade priorities and should include its recommendations in its fiscal 2001 budget request, which President Clinton will send to Congress early next year. ["Congress needs policy on shuttle, expert says," **Florida Today**, October 22, 1999, p 1A & 2A.]

OCTOBER 25: Space Shuttle Status Report, Monday, October 25, 1999. STS-103: Work in progress: Wiring protection efforts are nearing completion on board Shuttle Discovery. Recently completed wiring inspections revealed 57 incidents of exposed conductors. These damaged areas have been repaired and wiring protection methods have been implemented. Testing of the orbiter's repaired wiring system continues this week. Over the weekend, workers completed refilling activities on the right hand orbital maneuvering system (OMS) manifold No. 5. Ku band system testing and external tank door retests are also complete. Preparations are in work for payload bay door closure tomorrow. The Orbiter Rollout Review is scheduled this

Thursday and will determine Discovery's readiness to leave the Orbiter Processing Facility. Discovery is slated to move to the Vehicle Assembly Building on Nov. 1 to complete external tank mating activities. Space Shuttle Discovery is then expected to roll out to Launch Pad 39B on Sunday, Nov. 7. STS-99: Work in progress: Endeavour's wiring inspections are expected to conclude this week. To date, inspections have revealed 45 exposed conductors. Wiring protection and repair efforts continue. Auxiliary power unit wiring tests are in work and main engine controller verification is scheduled to occur this week. Work on Endeavour's nose and main landing gear is ongoing. STS-101: Work in progress: Wiring inspections on Atlantis are about 85 percent complete and have revealed a total of 34 exposed conductors to date. Wiring repair and protection efforts are ongoing. Checks on the water spray boiler are in work and Ku band system testing begins later this week. Inspections of Atlantis' right and left-hand inboard elevons revealed minor damage to mechanical actuators. The damaged parts will be replaced over the next several days. Installation of the right hand orbital maneuvering system pod will occur this week. Bruce Buckingham. (1999). **Kennedy Space Center Space Shuttle Status Report** [Online]. Available E-mail: domo@news.ksc.nasa.gov/subscribe shuttle-status [1999, October 25].]

OCTOBER 27: Future development at Kennedy Space Center will have little or no effect on access to Playalinda Beach, said an official with Spaceport Florida Authority. Ed O'Connor told the Titusville City Council that none of the five space vehicles being proposed for KSC would require the beach access road to be closed for lengthy periods. "There is no higher priority than preservation of beach access," O'Connor said of top-level discussions taking place at the space center. Playalinda Beach, which is part of Canaveral National Seashore, and the Merritt Island National Wildlife Refuge are on NASA-owned property. ["KSC growth won't affect beach access, Spaceport Flight official tells Titusville," **Florida Today**, p 1A.]

OCTOBER 29: Space Shuttle Status Report, Friday, October 29, 1999. STS-103: Work in progress: Shuttle managers will resume discussions on Monday at 1 p.m. to determine Discovery's readiness to leave the Orbiter Process Facility. The orbiter's planned transfer to the Vehicle Assembly Building will now occur no earlier than Monday evening. The postponement accommodates unplanned work to repair a temperature sensor on Discovery's No. 2 nitrogen tank, in the orbiter's midbody. Retests of Discovery's repaired and protected wiring are ongoing and will continue through the vertical processing flow. Several standard prelaunch tests will accommodate necessary functional and redundancy checks of the wiring. Once the orbiter, external tank and booster mating operations are complete, Space Shuttle Discovery will roll out to Launch Pad 39B. Rollout to the pad is currently slated for Sunday, Nov. 7. STS-99: Work in progress: Endeavour's wiring inspections, repair and protection installation continue. The orbiter payload premate test was completed Thursday. Tests on the auxiliary power unit wiring are in progress, as are Shuttle main engine controller verifications. Work on Endeavour's nose and main landing gear continues. STS-101: Work in progress: Shuttle managers announced today that the launch of Space Shuttle Atlantis on mission STS-101 will occur no earlier than March 16. The wiring inspections and repair efforts that remaining on the orbiter, along with the unplanned replacement of the ammonia boiler will require time to accommodate the Shuttle's processing needs. Inspections of Atlantis' ammonia boiler this week revealed corrosion, which lead to the replacement decision. Evaluation of the orbiter's damaged elevons continues. The damaged parts will be replaced over the next several days with no additional impact to the schedule. Installation of the right hand orbital maneuvering system pod occurs this week. Bruce Buckingham. (1999). **Kennedy Space Center Space Shuttle Status Report** [Online]. Available E-mail: domo@news.ksc.nasa.gov/subscribe shuttle-status [1999, October 29].]

◆ Technicians preparing space shuttle Atlantis for a flight next March have inadvertently damaged the ship's wing flaps, NASA said Friday. The technicians were testing the inboard elevons on Atlantis' wings late last week when they heard a noise, NASA spokesman Joel Wells said. They discovered the push rods on the flaps, or elevons, had been bent and some panels damaged. All of the damaged parts must be replaced. Although the mishap still is being investigated, NASA thinks the technicians failed to release small doors before moving the elevons, Wells said. That apparently caused the damage, he said. ["Atlantis' flaps damaged, probably by technicians," **The Orlando Sentinel**, October 30, 1999, p A-12.]

DURING OCTOBER: A launch services contract request for proposals (RFP) that would be worth more than $5 billion to winning contractors has been issued by the NASA Kennedy Space Center, Fla. The launch services RFP covers a broad range of planned NASA expendable launch vehicle needs across light, medium and heavy payloads. Nearly 70 major NASA unmanned launches through 2010 could eventually be part of several launch service contracts planned to be awarded under the program. The RFP has special provisions to enable new emerging and innovative launch service providers to compete for the contracts against established firms. Under a fast-track plan, corporate proposals are due at Kennedy by Jan. 7, 2000, and the NASA awards are to be made by June 2000. ["World News Roundup," **Aviation Week & Space Technology**, October 25, 1999, p 24.]

NOVEMBER

NOVEMBER 1: With a push of 10 buttons, Gov. Jeb Bush launched Brevard County's new 321 area code Monday with a call to Kennedy Space Center. KSC Director Roy Bridges was with Bush as he telephoned from the Public Service Commission in Tallahassee. At the other end of the line, in a conference room at the space center, about 25 NASA and local officials waited with Jim Jennings, KSC's deputy director for business operations, who answered the phone. For the next 11 months, callers will be able to use the new area code or the old one, 407. Brevard County is one of the smallest geographic regions in the state with its own code. ["Governor makes first 321 call to KSC officials," Florida Today, November 2, 1999, p 1B.]

◆ EG&G, Inc., now operates under a new name – PerkinElmer, Inc. The company is a former NASA base operations contractor in Brevard County. ["EG&G gets new name, new focus," Florida Today, November 2, 1999, p 10C.]

◆ AFL-CIO President John Sweeney met with workers, managers and labor officials Monday during a tour of Kennedy Space Center and Cape Canaveral Air Station. The American Federation of Labor-Congress of Industrial Organizations, which has 13 million members, is affiliated with more than 5,000 union members at KSC, Cape Canaveral Air Station and Patrick Air Force Base. Chris Hunt, vice president of Transport Workers Union of America Local 525 in Cocoa Beach, which represents about 2,000 contracted workers at KSC, called Sweeney's visit "historic." ["AFL-CIO chief visits KSC," Florida Today, November 2, 1999, p 10C & 9C.]

◆ For the second time this year, an Internet auction site has posted a listing for a reported piece of wreckage from the space shuttle Challenger disaster. Federal law prohibits anyone from owning challenger debris, and officials at eBay of San Jose, Calif., said Monday they were investigating the listing for a thermal tile from the space shuttle. Challenger exploded a little more than a minute into flight Jan. 28, 1986. Anyone possessing wreckage could face up to 10 years in prison and fines up to $10,000. The same site removed an ad for an "authentic Challenger o-ring" in January. "We are talking with NASA at this point and if it is indeed an illegal item, we will remove it" from our listings, said Kevin Pursglove, an eBay spokesman. In the meantime, NASA has turned over the matter to its security office and could take steps to reclaim the tile if it is from Challenger, agency spokeswoman Kirsten Williams said in Washington, D. C. The tile's owner claims to have been on the first ship that arrived at the wreckage site. ["Auction Web site posts 2nd Challenger piece," Florida Today, November 2, 1999, p 1A.]

NOVEMBER 2: Space Shuttle Status Report, Tuesday November 2, 1999. STS-103: Work in progress: Yesterday, Shuttle managers concluded the orbiter rollout review for Shuttle Discovery. However, the decision to proceed with prelaunch processing for mission STS-103 will not be made until after a preflight readiness review which starts today at 2 p.m. Top Shuttle managers are assembling for the meeting at Johnson Space Center to conduct a thorough review of Shuttle wiring assessments and corrective actions taken before transferring Discovery out of the Orbiter Processing Facility. Wiring repair and protection efforts are complete and the payload bay doors are closed. Repairs on Discovery's No. 2 nitrogen tank are complete and tests indicate that the temperature sensor is in working order. Aft compartment close-outs and structural leak tests are complete and weight and center of gravity tests were conducted yesterday. Preparations are under way to support Discovery's rollover to the Vehicle Assembly Building as early as 10 a.m. tomorrow, but a decision to roll will not be made until after the Shuttle wiring review. Bruce Buckingham. (1999). **Kennedy Space Center Space Shuttle Status Report** [Online]. Available E-mail: domo@news.ksc.nasa.gov/subscribe shuttle-status [1999, November 2].]

NOVEMBER 3: Space Shuttle Status Report, Wednesday, November 3, 1999. STS-103: Work in progress: Orbiter Discovery is currently slated to roll from OPF bay 1 tomorrow at about 10 a.m. to be mated with its external tank and solid rocket boosters in the Vehicle Assembly Building. This delay, along with plans to replace main engine No. 3, have resulted in managers assessing the remainder of Discovery's processing schedule. Preliminary plans still support a launch date in early December, although a specific target won't be selected for several days. Managers discussed an issue with engine No. 3 during their meeting. A broken drill bit measuring one-half inch in length and weighing less than one-half gram is located in a coolant cavity in the right engine, serial number 2045. The bit was broken during routine processing of the engine at KSC. KSC managers are developing a work schedule to replace engine No. 3 at the launch pad. Replacement efforts will take about 10 days with most of the work conducted in parallel with other launch processing activities. Bruce Buckingham. (1999). **Kennedy Space Center Space Shuttle Status Report** [Online]. Available E-mail: domo@news.ksc.nasa.gov/subscribe shuttle-status [1999, November 3].]

NOVEMBER 4: Space Shuttle Status Report, Thursday, November 4, 1999. STS-103: Work in progress: Orbiter Discovery rolled out of OPF bay 1 today at about 10 a.m. Later tonight, workers will begin efforts to mate the orbiter to the external tank and twin solid rocket boosters located in VAB high bay 1. Current work plans show the Space Shuttle rolling out to Launch Pad 39B early next week. Once at the pad, preparations to replace Shuttle main engine No. 3 will begin. Managers expect that effort to take about 10 days, with most of the work done in parallel with other prelaunch work. Shuttle managers are assessing the remainder of Discovery's processing schedule and expect to announce a new target launch date next week. Bruce Buckingham. (1999). **Kennedy Space Center Space Shuttle Status Report** [Online]. Available E-mail: domo@news.ksc.nasa.gov/subscribe shuttle-status [1999, November 4].]

NOVEMBER 5: Space Shuttle Status Report, Friday, November 5, 1999. STS-103: Work in progress: Orbiter Discovery has been mated to the external tank and solid rocket boosters stack in VAB high bay 1. Electrical and mechanical connections will conclude late Friday and the orbiter/external tank umbilical mate occurs Saturday. Over the weekend, workers will begin preparations for the engine No. 3 replacement work scheduled to happen at the launch pad. Managers currently plan to transfer Discovery out to Launch Pad 39B Tuesday, Nov. 9 beginning at about 2 a.m. Payload managers plan to transfer the Hubble Servicing Mission cargo to the launch pad Monday, Nov. 8 with installation into the orbiter slated for Nov.12. KSC managers are developing the remainder of Discovery's prelaunch processing schedule and expect to brief Shuttle program managers at a meeting Monday morning. Program managers are likely to set a new target launch date for STS-103 after the meeting. Bruce Buckingham. (1999). **Kennedy Space Center Space Shuttle Status Report** [Online]. Available E-mail: domo@news.ksc.nasa.gov/subscribe shuttle-status [1999, November 5].]

NOVEMBER 6: At 6 feet tall, the rocket will be shorter than your average NBA player. And it has to hoist only about 9 pounds of cargo into space. But this vehicle is giving Kennedy Space Center engineers a challenge and a chance to extend their launch expertise into the solar system by developing a rocket to blast off from Mars. Its assignment: Carry Martian dust and rocks into orbit around the planet so another spaceship can retrieve them for delivery to Earth in 2008. The KSC-led team knows the rocket project has little room for error. "It's a relatively small vehicle, but it has to do some pretty hard things," said Dave Taylor, KSC project engineer for the Mars rocket. Given the task of overseeing the project in September, the KSC team in December plans to choose the company that will build the rocket. The company will have to begin work immediately, so two rockets can be ready for flights to Mars on a pair of missions scheduled for launch in 2003 and 2005. KSC officials say the rocket itself is not hard to design, but getting it to work smoothly from an alien launch site 400 million miles away won't be easy. Beyond its importance to the Mars program, the new rocket is crucial to KSC if the center is to add new talents to its resume, said Roy Bridges, KSC's director and a former astronaut. He said the goal is to make KSC a place where future space technologies are born. ["KSC designs Mars-bound rocket," Florida Today, November 7, 1999, p 1A & 2A.]

NOVEMBER 8: Space Shuttle Status Report, Monday, November 8, 1999. STS-103: Work in progress: The planned rollout of Space Shuttle Discovery to Launch Pad 39B has been delayed beyond Tuesday, Nov. 9. During standard Shuttle Interface Test activities this morning, engineers in firing room No. 1 noted loss of command capability for a range safety cable that supports both solid rocket boosters. Subsequent visual inspections revealed damage to the SRB cross-strap cable that runs between the external tank (ET)/SRB attach points and through the ET intertank. KSC managers have decided to replace the cable and are developing a plan to accommodate the unexpected work. The impact of this work on the remainder of Discovery's processing schedule is being assessed. Preliminary reports indicate that an early December launch date is still achievable. The Hubble Servicing Mission cargo was transported to the launch pad early this morning and transfer into the payload changeout room is in work. Bruce Buckingham. (1999). **Kennedy Space Center Space Shuttle Status Report** [Online]. Available E-mail: domo@news.ksc.nasa.gov/subscribe shuttle-status [1999, November 8].]

NOVEMBER 9: Space Shuttle Status Report, Tuesday, November 9, 1999. STS-103: Work in progress: Space Shuttle Discovery is scheduled to roll out to Launch Pad 39B on Saturday, Nov. 13, with first motion slated for 2 a.m. The remainder of Discovery's processing schedule leads to a target launch date of Dec. 6. Workers have already removed the damaged range safety cable and replacement efforts begin later today. Engineers plan to retest the range safety system Wednesday and close-out the work area Thursday. The damaged range safety cable relays a redundant emergency destruction signal between the solid rocket boosters (SRB) in the unlikely event of a contingency. The cable being replaced runs from the right-hand SRB forward attach point, through the external tank and connects to the left-hand booster. The cable was damaged during close-out operations causing yesterday's test failure. Since the orbiter will remain in the VAB for cable replacement, Shuttle managers have decided to replace Discovery's main engine No. 3 in the VAB as well. Engine replacement efforts are in work and will conclude Thursday. Engine close-outs and verifications will be performed at the pad. The Hubble Servicing Mission cargo has been transferred into the payload change-out room at Launch Pad 39B and installation into the orbiter's payload bay is slated for Nov. 16.

Major Processing Milestones:

Shuttle to the Launch Pad	(Nov. 13)
Payload installed into orbiter	(Nov. 16)
Terminal Countdown Demonstration Test	
(crew dress rehearsal)	(Nov. 17)
Shuttle main engine test	(Nov. 23)

Bruce Buckingham. (1999). **Kennedy Space Center Space Shuttle Status Report** [Online]. Available E-mail: domo@news.ksc.nasa.gov/subscribe shuttle-status [1999, November 9].]

NOVEMBER 11: United Space Alliance this week was honored with the 1999 Florida Civil Rights Corporate Advocacy Award by the Florida Commission on Human Relations. The award, which was presented at the 9th Annual Civil Rights Conference in Miami, recognizes individuals and organizations who have significantly contributed to the advancement of fair treatment and equal opportunity in Florida. According to the nomination, USA was recognized for demonstrating exceptional results in preventing the formation of barriers that challenge workforce diversity, and for developing exemplary and novel practices to ensure opportunities for all employees to advance. ["USA recognized for diversity programs," **USA News Release,** November 11, 1999.]

NOVEMBER 13: Space shuttle Discovery is sitting on its launch pad today as NASA begins final preparations for next month's flight to the orbiting Hubble Space Telescope. Now that wiring problems are fixed and a new main engine is installed,

Discovery is set to fly Dec. 6. The mission: to repair Hubble's aiming system. The seven Discovery astronauts are expected to arrive at Kennedy Space Center today from Houston. During the next several days, the crew will receive emergency training and participate in a practice countdown. Discovery made a seven-hour, 4.2 mile trip from KSC's Vehicle Assembly Building to launch pad 39B on Saturday morning, about two months behind schedule because of technical problems plaguing the shuttle program. ["Shuttle on pad after series of problems, delays," <u>Florida Today</u>, November 14, 1999, p 1A.]

◆ From cellular phone satellites to the laptops that astronauts use aboard NASA's space shuttles, the aerospace industry claims it is ready for the year 2000. "We are cautiously optimistic that we have covered everything," said Brian Dunbar, a NASA spokesman at the agency's headquarters in Washington, D.C. "But there are never any guarantees with anything as unprecedented as this." The National Aeronautics and Space Administration started working in earnest on Y2K in August 1996. Now, more than three years and $70 million later, agency officials say critical agency systems are ready. "We have been incredibly thorough," said Art Beller, year 2000 project manager at Kennedy Space center. "There isn't a major system or a minor system that hasn't been touched. The work included checking everything from the elevators inside KSC buildings to the critical computer software that tracks the health of a space shuttle as the seconds tick down to liftoff. NASA will not allow any of its shuttles to be in orbit when the calendar rolls into the year 2000. But that's nothing new. Traditionally, the last two weeks of the year are "down" periods for NASA. Non-critical systems are turned off, and annual maintenance is done on buildings and grounds. The tradition could be infringed a little this year, with agency officials saying they will launch shuttle Discovery on a 10-day repair mission to the Hubble Space Telescope as late as Dec. 14. They will fly no later, however. Allowing for weather delays to the landing, the agency does not want to risk flying over the New Year holiday. ["NASA officials say everything is ready for the big day," <u>Florida Today</u>, November 14, 1999, p 4A.]

NOVEMBER 15: Space Shuttle Status Report, Monday, November 15, 1999. STS-103: Work in progress: Space Shuttle Discovery rolled out to Launch Pad 39B on Saturday, Nov. 13. First motion from the Vehicle Assembly Building began a 7:27 a.m. and the Shuttle was hard down at the pad by 2:17 p.m. Discovery had been scheduled to begin its transfer to the pad at 2 a.m., but inspections of minor external tank foam damage delayed the departure. While Discovery was in the VAB, workers completed efforts to replace Shuttle main engine No. 3 and the solid rocket booster range safety cable. The Shuttle Interface Test validated all Space Shuttle connections prior to rollout. Final checks of the replaced main engine will be completed at the pad. Launch pad validations are under way and installation of the payload into Discovery's payload bay is slated for tomorrow. The seven-member flight crew arrived at KSC Sunday afternoon and will participate in Terminal Countdown Demonstration Test activities through Wednesday. The launch day dress rehearsal concludes Wednesday at 11 a.m.

with a simulated main engine cutoff. Shuttle managers will convene the STS-103 Flight Readiness Review on Friday, Nov. 19 to discuss the overall readiness of all Space Shuttle systems for flight.

Major Processing Milestones (targets only):

Payload installed into orbiter   (Nov. 16)
Terminal Countdown Demonstration Test
(crew dress rehearsal)   (Nov. 17)
Flight Readiness Review   (Nov. 19)
Shuttle main engine test   (Nov. 23)

STS-99: Work in progress: Wiring retests on orbiter Endeavour are ongoing and midbody closeouts conclude today. Installation of the SRTM payload into Endeavour's payload bay is scheduled to occur tomorrow.

Major Processing Milestones (targets only):

Payload installed into   (Nov. 16)
Orbiter transferred to VAB   (Dec. 2)
Space Shuttle rolls out to Launch Pad 39A (Dec. 8)

Bruce Buckingham. (1999). **Kennedy Space Center Space Shuttle Status Report** [Online]. Available E-mail: domo@news.ksc.nasa.gov/subscribe shuttle-status [1999, November 15].]

NOVEMBER 16: An expensive makeover has given the Kennedy Space Center's Visitor Complex a sleek new look that includes a high-tech conference center and an exhibit on early space exploration scheduled to open in the next few weeks. The final piece of the $120 million project – an exhibit on space exploration in the next millennium – will be ready soon afterward. The highlight of Early Space Exploration – which focuses on the Mercury and Gemini programs – arguably begins in the queuing area, where visitors get background from vintage televisions in a Kennedy-era kitchen and living room filled with '60s kitsch. From there, they go on to view the actual Mercury Mission Control Room. "The only thing we had to replace was the tiles on the floor," said visitor complex spokeswoman Amy Maguire. "All the consoles have ashtrays except the physician's because he knew better." The adjoining Kurt Debus Conference Facility is all modern, though. The conference center, named for the first director of the Kennedy Space Center, has full audio-visual capabilities and seats 400 or serves 700 buffet style. A glass-walled rotunda overlooks the refurbished Rocket Garden, one of the visitor center's signature attractions. The $120 million renovation, which began in 1996 with the opening of the Apollo/Saturn V Center, marks the first improvements in recent history for the complex as well as the most tangible results of new management. In 1995, Delaware North Parks Services of Spaceport Inc. won the NASA contract to operate the visitor center, which had been run by TW Recreational Services for nearly 20 years. ["Attraction launches space-age makeover," **The Orlando Sentinel**, November 17, 1999, p B-1 & B-6.]

◆ Just days after NASA's Hubble Space Telescope closed its eye on the cosmos, the astronauts who will perform a repair mission next month say their flight will be more complicated than they originally thought. Hubble's precision pointing system failed when the fourth of six devices called gyroscopes malfunctioned Saturday, halting its science work. The telescope needs three of the units working to observe the universe. Because of the pointing failure, the new challenge added to Discovery's mission is the uncertainly of how the telescope will be positioned when the shuttle arrives two days after its planned Dec. 6 launch. That concern will require precise flying more than 300 miles above Earth by commander Curt Brown, a veteran of five shuttle flights. Brown and the six other crew members discussed the mission Tuesday at Kennedy Space Center, where they are participating in a practice countdown today. ["Discovery's mission gets trickier," Florida Today, November 17, 1999, p 1A.]

◆ With a practice countdown completed, NASA officials will hold a final review Friday to firm up the launch date for next month's shuttle mission to service the Hubble Space Telescope. The space agency hopes Discovery will be ready for launch at 2:37 a.m. on Dec. 6. "That's our target and that's what we are working toward," NASA spokesman Lisa Malone said. Discovery's seven astronauts were aboard their ship Wednesday morning for the final three hours of simulated launch countdown. Later, the crew returned to their home base at Johnson Space Center in Houston to wrap up training. They are due back at Kennedy Space Center on Dec. 2. ["NASA firms up launch date for Discovery," Florida Today, November 18, 1999, p 5A.]

NOVEMBER 18: The STS-103 crew finished their Terminal Countdown Demonstration Test activities yesterday with training on emergency egress from the orbiter plus inspection of the Hubble servicing cargo in Space Shuttle Discovery's payload bay. They were scheduled to leave KSC in their T-38 jets about 3 p.m. today. Shuttle managers will convene the STS-103 Flight Readiness Review tomorrow to discuss the overall readiness of all Space Shuttle systems for flight. ["TCDT complete, crew on STS-103 mission ready for launch Dec. 6," KSC Countdown, November 18, 1999.]

◆ NASA has found more damaged electrical wires on shuttle Discovery, officials said Thursday, prompting new inspections of the ship as work continues toward a Dec. 6 launch to repair the Hubble Space Telescope. Officials don't know whether the checks and repairs will delay the liftoff, said Kennedy Space Center spokesman Joel Wells. All the work can be done while the shuttle sits on its launch pad, he said. The new damage, described as minor, was found on wiring between the spaceship, its external tank and solid rocket boosters. Originally scheduled for October, the flight was postponed when NASA ordered wiring inspections on its entire fleet after shuttle Columbia experienced a troublesome short circuit during a July launch. "The important thing here is that we are operating in an environment of heightened awareness over wiring issues," Wells said. "We are being very meticulous in our inspections." When launched, the ship is to carry seven astronauts to repair Hubble's precision-pointing system. ["Damaged wires found on shuttle," Florida Today, November 19, 1999, p 1A.]

NOVEMBER 19: Space Shuttle Status Report, Friday, November 19, 1999. STS-103: Following today's Flight Readiness Review at KSC, managers decided to keep the target launch date of no earlier than Dec. 6 for mission STS-103. Managers plan to convene early next week to further address potential changes to Discovery's target launch date. Also, a follow-up Flight Readiness Review will be held after the Thanksgiving Holiday weekend to permit managers the opportunity to re-address outstanding engineering issues and to establish an official launch date. Bruce Buckingham. (1999). **Kennedy Space Center Space Shuttle Status Report** [Online]. Available E-mail: domo@news.ksc.nasa.gov/subscribe shuttle-status [1999, November 19].]

NOVEMBER 22: Space Shuttle Status Report, Monday, November 22, 1999. STS-103: Work in progress: Shuttle managers today set December 9 as the launch date for Discovery on STS-103. The adjusted launch date allows technicians time to repair minor electrical wiring damage that was found recently in an umbilical between the Shuttle orbiter and the external tank. The schedule also allows the Shuttle workforce to observe the Thanksgiving holidays. Managers plan to reconvene the STS-103 Flight Readiness Review on Dec. 1 at 2 p.m. to obtain a final status on the recent work completed on Discovery. Friday, workers at Launch Pad 39B completed payload interface verification testing and today are conducting the planned end-to-end test. Payload bay close-outs begin this week and Discovery's payload bay doors will be closed for flight Nov. 24. Over the weekend, oxidizer and fuel reactants were loaded into Discovery's onboard storage tanks. Leak checks on Shuttle main engine No. 3 are complete and good. Routine inspections of Discovery's aft compartment are in work in preparation for next week's orbiter aft close-outs.

Major Processing Milestones (targets only):

Shuttle main engine test	(Nov. 23)
Payload bay doors closed for flight	(Nov. 24)
Shuttle Interface Test repeated	(Nov. 29)
Ordnance installation	(Dec. 1)
Orbiter aft compartment closed-out	(Dec. 4)
Launch countdown begins	(Dec. 6)

STS-99: Work in progress: The Shuttle Radar Topography Mission payload was installed inside Endeavour's payload bay last week and orbiter interface verification testing has been completed. The payload bay doors are slated for closure this week. Managers plan to roll Endeavour to the Vehicle Assembly Building Dec. 2 to be mated to the external tank and solid rocket boosters. Once those activities are complete, Shuttle Endeavour will roll out to Launch pad 39A.

Major Processing Milestones (targets only):

Payload bay doors closed for	(Nov. 28)
Orbiter transferred to VAB	(Dec. 2)
Shuttle rolls to Launch Pad 39A	(Dec. 7)

Bruce Buckingham. (1999). **Kennedy Space Center Space Shuttle Status Report** [Online]. Available E-mail: domo@news.ksc.nasa.gov/subscribe shuttle-status [1999, November 22].]

◆ The Navy has a new satellite in orbit today where it's ready to join a network of pentagon spacecraft linking warships to shore. The $200 million communications satellite was carried into space Monday on a Lockheed Martin Atlas 2A rocket that lifted off at 11:06 p.m. from Cape Canaveral Air Station. ["Rocket reaches orbit," **Florida Today**, November 23, 1999, p 1A.]

NOVEMBER 29: Space Shuttle Status Report, Monday, November 29, 1999. STS-103: Work in progress: Last Wednesday, Shuttle workers closed Discovery's payload bay doors and completed planned main engine testing. Repair efforts on Discovery's umbilical wiring are complete and validation tests begin tonight. Following the Thanksgiving holiday, workers resumed orbiter electrical wiring inspections in Discovery's aft compartment. Today, engineers will conduct the helium signature leak test. Ordnance installation is slated to occur midweek and replacement of a leaky quick disconnect on auxiliary power unit No. 2 is planned for Wednesday evening. Aft engine compartment close-outs are expected to conclude late Sunday. Workers are replacing Discovery's mass memory unit No. 1 today. Wednesday afternoon, Shuttle managers will reconvene a follow-up Flight Readiness Review to obtain a final status on prelaunch preparations. The flight crew is scheduled to arrive at KSC's Shuttle Landing Facility Saturday night, and the launch countdown begins Monday, Dec. 6 at 5:30 a.m.

Major Processing Milestones (targets only):

Ordnance installation	(Dec. 1)
Flight crew arrival at	(Dec. 4 at about 8 p.m.)
Orbiter aft compartment close-outs complete	(Dec. 5)
Launch countdown begins	(Dec. 6 at 5:30 a.m.)

STS-99: Work in progress: Workers began minor repair efforts on Endeavour's umbilical wiring harness on Sunday, and wiring inspections continue in the orbiter's aft engine compartment. Orbiter Endeavour's midbody close-outs are ongoing as workers prepare to close the orbiter's payload bay doors for flight later tonight. Managers plan to transfer Endeavour to the Vehicle Assembly Building on Thursday to be mated to the external tank and solid rocket boosters located in VAB high bay 1. Endeavour's Orbiter Rollout Review is scheduled for Tuesday morning.

Major Processing Milestones (targets only):

Payload bay doors closed for flight	(Nov. 29)
Orbiter transferred to VAB	(Dec. 2)
Shuttle rolls to Launch Pad 39A	(Dec. 7)

Bruce Buckingham. (1999). **Kennedy Space Center Space Shuttle Status Report** [Online]. Available E-mail: domo@news.ksc.nasa.gov/subscribe shuttle-status [1999, November 29].]

DURING NOVEMBER: Lockheed Martin has completed one of the most extensive launch pad modifications in the history of Cape Canaveral to convert Complex 36B here for Atlas III flights. A major challenge for the Atlas team was modifying Pad B while also keeping it open for Atlas Centaur missions, said Adrian A. Laffitte, director of Atlas programs for Lockheed martin. The adjoining Complex 36A Atlas Centaur pad is being retained in its Atlas IIAS configuration. The $8 million upgrade to Pad B included 66 major modifications to its 4-million-lb., 250-ft. tall mobile service tower (MST) and the overall pad. Many of these changes have been to increase the MST height for the taller Atlas II, while also decreasing its weight and wind cross section. ["Major Pad 36B Mods Support Atlas III," **Aviation Week & Space Technology**, November 22, 1999, p 55.]

DECEMBER

DECEMBER 1: A recent string of multibillion-dollar U.S. space launch failures can be traced to flawed workmanship and engineering by the contractors who built the Titan 4 and Delta 3 rockets, a Pentagon study concludes. The report released Wednesday also cited fragmented lines of authority within the government for overseeing the space launch program. The study was ordered by the Air Force after the malfuctioning of a Delta 3 rocket in May left a commercial communication satellite in a useless lopsided orbit – the fifth failed space launch since August 1998. The May failure was the second botched flight in a row for the Delta 3 built by Boeing Co. ["Study faults contractors, government alike for string of launch failures," **Florida Today**, December 2, 1999, p 1A & 2A.]

◆ Space Shuttle Status Report, Wednesday, December 1, 1999. STS-103: Work in progress: Space Shuttle managers today completed a review of Shuttle Discovery's readiness for flight on mission STS-103. They maintained a launch date of Dec. 9 for Discovery, although a further review of work remaining to close-out Discovery for flight will be held later this week. The launch date may be adjusted slightly pending the status of remaining work that is reported at that time. At the launch pad, workers completed replacement of Discovery's mass memory unit No. 1. Close-out work on the recently repaired orbiter/external tank umbilical wiring harness continues as engineers conduct validation tests on that system. Tonight, workers are slated to begin Shuttle ordnance installation and replacement of a leaky quick disconnect on auxiliary power unit No. 2 is planned for later this week. Routine orbiter aft compartment close-outs continue along with wiring inspections. STS-99: Work in progress: Workers in the OPF have completed wiring close-outs on Orbiter Endeavour and the payload bay doors were closed last night. Preparations are under way for Endeavour to roll to the Vehicle Assembly Building Thursday at about 10 a.m. The orbiter will be mated to the external tank and solid rocket boosters in VAB high bay 1Thursday night and managers plan to transfer the Space Shuttle to Launch Pad 39A Dec. 7. Bruce Buckingham. (1999). **Kennedy Space Center Space Shuttle Status Report** [Online]. Available E-mail: domo@news.ksc.nasa.gov/subscribe shuttle-status [1999, December 1].]

DECEMBER 2: NASA managers today set Saturday, Dec. 11, 1999, as the launch date for NASA's final Space Shuttle mission this century. The 96th Space Shuttle mission will be highlighted by four space walks to service the Hubble Space Telescope. Discovery is scheduled to lift off from Launch Pad 39-B at NASA's Kennedy Space Center, FL, at 12:13 a.m. EST, the opening of a 38-minute launch window. Discovery's planned 10-day flight, designated Shuttle mission STS-103, is scheduled to end with a night landing at Kennedy at about 9:21 p.m. EST on Dec. 20. ["Launch of Hubble servicing mission set for December 11," **NASA News Release #99-141**, December 2, 1999.]

DECEMBER 3: Space Shuttle Status Report, Friday, December 3, 1999. STS-103: Work in progress: Shuttle managers determined today that an additional inspection of umbilical wiring is required on Shuttle Discovery. The electrical wire being inspected supports the pyrotechnic initiator controller for Discovery's left-hand solid rocket booster and is part of the Shuttle's liquid oxygen umbilical assembly. Recently, workers were tasked to inspect and repair minor insulation flaws on the wires located in the orbiter umbilical harnesses. The unplanned work concluded on Wednesday, Dec. 1. This latest inspection will determine if additional work is required to ensure the flight readiness of the single pyrotechnic wire. STS-99: Work in progress: Yesterday afternoon, orbiter Endeavour arrived in the Vehicle Assembly Building and preparations began to mate the orbiter to the external tank and solid rocket boosters in high bay 1. Today orbiter/external tank mating activities began and the Shuttle Interface Test begins early Monday morning. Shuttle managers today decided to replace Endeavour's main engine No. 3 while the Shuttle is in the VAB. This additional work will delay the Shuttle's roll out to Launch Pad 39A until Dec. 13. Analysis of a separate test engine revealed delamination on the wall of the engine's main combustion chamber following routine testing procedures. Further data gathering revealed that one of Endeavour's engines had undergone similar testing procedures and as a precaution managers opted to replace the suspect engine. Bruce Buckingham. (1999). **Kennedy Space Center Space Shuttle Status Report** [Online]. Available E-mail: domo@news.ksc.nasa.gov/subscribe shuttle-status [1999, December 3].]

DECEMBER 6: NASA's mission to repair the ailing Hubble Space Telescope may have to wait until at least Saturday night to begin after another faulty electrical wire was found aboard shuttle Discovery. The problem wire, found Monday inside Discovery's rear engine compartment, is part of the network of cables attached to one of the shuttle's three main engines. Final pre-launch inspections revealed the wire had about one-eighth of an inch of exposed conductor, and managers ordered the wire replaced, KSC spokesman Joel Wells said. Once the wire is replaced, the new cable must be tested. "Managers haven't confirmed Discovery's launch date, but a one-day slip in the launch is possible," Wells said. The mission was first targeted for launch Oct. 14. Although a new launch date won't be announced until the work is complete, it will probably be 11:42 p.m. Saturday at the earliest before Discovery can take off. The previous launch window was from 12:13 to 12:51 a.m. Saturday. ["Wire problem likely to delay shuttle launch," **Florida Today**, December 7, 1999, p 1A.]

◆ Space Shuttle Status Report, Monday, December 6, 1999. STS-103: Work in progress: The seven-member STS-103 flight crew arrived at KSC's Shuttle Landing Facility tonight at about 8:20 p.m. During the days leading up to launch, the crew will participate in orbiter and mission familiarization activities. They will take opportunities to fly in the Shuttle Training Aircraft and will undergo routine preflight medical exams. At Launch Pad 39B, workers continue efforts to close out Shuttle Discovery's aft compartment for flight. During routine engine compartment inspections today, workers found a 1/8-inch nick in the insulation of a Shuttle main engine wire. The wire provides command and feedback support for Discovery's No. 2 main engine. Following an engineering evaluation meeting this evening, Shuttle managers decided to replace the damaged wire and reported that a one-day launch slip is possible. Managers will further assess the impact to the processing schedule and expect more information by late tomorrow. Bruce Buckingham. (1999). **Kennedy Space Center Space Shuttle Status Report** [Online]. Available E-mail: domo@news.ksc.nasa.gov/subscribe shuttle-status [1999, December 6].]

DECEMBER 7: Space Shuttle Status Report, Tuesday, December 7, 1999. STS-103: Work in progress: This afternoon, Shuttle managers moved the launch of Shuttle Discovery to Dec. 11 at 11:42 p.m. The delay accommodates additional wiring work in the orbiter's aft compartment. Today, technicians are completing efforts to replace and retest a wiring harness that provides command and feedback support to the main engine No. 2 controller. One wire in the harness had a 1/8-inch nick in its Teflon insulation, and engineers decided last night to replace the entire harness. Engine No. 2 will undergo flight readiness tests, and leak checks will be repeated. Aft compartment close-outs are expected to conclude early Thursday, and the launch countdown is set to begin at 4 a.m. Thursday. The flight crew will take advantage of the additional time to review mission plans and spend time with family members. Bruce Buckingham. (1999). **Kennedy Space Center Space Shuttle Status Report** [Online]. Available E-mail: domo@news.ksc.nasa.gov/subscribe shuttle-status [1999, December 7].]

DECEMBER 8: Space Shuttle Status Report, Wednesday, December 8, 1999. STS-103: Work in progress: Shuttle managers have decided to delay the start of the STS-103 launch countdown by at least 24 hours while engineers evaluate a dented main propulsion system hydrogen line found during closeout inspections of Discovery's engine compartment. A final decision on whether or not the line must be replaced is expected tomorrow. If the line requires replacement, it is anticipated that the work would postpone launch by a minimum of several days. The four-inch diameter line carries liquid hydrogen fuel for the Space Shuttle main engines. Bruce Buckingham. (1999). **Kennedy Space Center Space Shuttle Status Report** [Online]. Available E-mail: domo@news.ksc.nasa.gov/subscribe shuttle-status [1999, December 8].]

DECEMBER 9: U.S. and Russian astronauts working on the international space station should limit their exposure to radiation during intense solar activity, a scientific advisory panel recommended Thursday. Plans for construction of the space station by 2004 call for some 43 space shuttle missions and about 1,500 space walks. The work will coincide with the peak of the 11-year cycle of solar activity next year, the National Research Council noted in its report. The council is an arm of the National Academy of Science, which provides scientific advice to government agencies. ["Scientific advisory panel urges radiation limits for astronauts," <u>Florida Today</u>, December 10, 1999, p 8A.]

◆ Mission managers decided Thursday to target a new launch date – Dec. 16 – for Discovery's 10-day flight to repair the Hubble Space Telescope. That's after engineers came up with a plan to quickly replace a dented fuel line that initially had threatened to bump the mission into January. The line circulates super-cold liquid hydrogen between the shuttle's external tank and its three main engines. If all goes according to plan, Discovery will have three possible launch dates – Dec. 16, 17 and 18 – before the flight would have to be postponed until next year. The launch window on Dec. 16 would extend from 9:18 to 9:59 p.m. "We will not have a very good idea of how close we are to making the 16th until Monday morning," said Ron Dittemore, the shuttle program manager at NASA's Johnson Space Center. "There's a number of milestones that have to be accomplished this weekend, since we've never done this particular change-out." Shuttle managers want Discovery back on the ground no later than Dec. 28 to allow plenty of time for servicing the orbiter and shutting down ground computers before any potential Y2K problems appear Jan. 1. Allowing two contingency days for possible bad weather, that means landing will be scheduled for Dec. 26. ["Latest shuttle launch plan calls for Christmas in orbit," The Orlando Sentinel, December 10, 1999, p A-6.]

DECEMBER 11: It's still uncertain whether shuttle Discovery will lift off this week on a mission to restore NASA's Hubble Space Telescope to working condition. To NASA, it represents much more than a visit to the observatory. After two years with few flights, Discovery's voyage will signal the start of a demanding launch schedule that must be met to build the $60 billion International Space Station during the next five years. The increase to about seven flights a year is not new to NASA, which had that many or more missions during most of the 1990s. But the renewed push will fall for the first time on a severely reduced work force at Kennedy Space Center, where the number of shuttle workers has been cut by 35 percent since 1995. Many experienced technicians have been lost in the cutbacks, causing concern the smaller work force could be strained by the high demands and flight safety put at risk. That's why NASA and independent experts say they'll be watching closely to make sure safety isn't jeopardized. ["Smaller work force faces tough launch demands," Florida Today, December 12, 1999, p 1A & 2A.]

DECEMBER 13: Union members from the Kennedy Space Center Visitor Complex have voted to accept a contract offer from Delaware North Park Services of Spaceport, averting the chance of a strike. Members of the International Association of Machinists Local 773 accepted the contract by a vote margin last week of 53 percent to 47 percent, said Scott Pendergrass, the union's business representative. There are about 300 workers in the union who staff the Visitor Complex, NASA's tourist attraction. The workers had rejected a previous contract offer and were preparing to go on strike. The new three-year contract, retroactive to Nov. 1, gives most workers average wage increases of 3 percent a year. The contract also includes a bidding system for job assignments, Pendergrass said. ["Strike averted at KSC," Florida Today, December 14, 1999, p 10C.]

DECEMBER 14: Space Shuttle Status Report, Tuesday, December 14, 1999. STS-103: Work in progress: The launch countdown for STS-103 began on schedule today at 1:30 a.m. Yesterday, workers completed aft compartment close-outs and Discovery's aft confidence test concluded last night. Current weather forecasts call for scattered to broken clouds at 3,000 feet and scattered clouds at 25,000 feet; visibility of 7 miles; winds from the north at 12 peaking to 20 knots; temperature at 52 degrees F; relative humidity at 69 percent; and no chance of showers. Weather officials indicate an 80 percent chance of favorable weather conditions for launch attempts on Thursday or Friday. By Saturday, the forecast declines to a 70 percent chance of favorable weather. Processing Milestones:
Cryogenics loaded in Discovery's onboard storage tanks   (Dec. 15)
Rotating Service Structure moved to park position   (Dec. 16 at 3:45 a.m.)
STS-99: Work in progress: Space Shuttle Endeavour arrived at Launch Pad 39A yesterday at about 12:30 p.m. Launch pad validations are under way and an auxiliary power unit hot fire test is scheduled today. Bruce Buckingham. (1999). **Kennedy Space Center Space Shuttle Status Report** [Online]. Available E-mail: domo@news.ksc.nasa.gov/subscribe shuttle-status [1999, December 14].]

◆   Technicians at Kennedy Space Center replaced a damaged fuel line on shuttle Discovery over the weekend, eliminating a potential obstacle to Thursday's planned launch. Discovery is scheduled to lift off on a 10-day mission to repair the Hubble Space Telescope during a launch window extending from 9:18 to 9:59 p.m. ["New fuel line may end Discovery's woes," <u>The Orlando Sentinel</u>, December 14, 1999, p A-9.]

DECEMBER 15: Space Shuttle Status Report, Wednesday, December 15, 1999. STS-103: Work in progress: Preparations for the launch of Space Shuttle Discovery continue on schedule today. Last night, work began to load the orbiter's fuel cell storage tanks with cryogenic reactants and loading operations concluded this morning. Discovery's main engines have undergone routine prelaunch preparations today and orbiter communication system activation occurs tonight. Flight crew equipment late stowage begins later this evening. Today, Shuttle managers and engineers have evaluated a welding issue involving external tank pressure lines. While the issue was identified on hardware intended for use on a future flight, engineers are confirming that the welding flaw was an isolated occurrence and that the pressure lines on Shuttle Discovery's external tank are not affected. With the weld analysis all but complete, managers expect to close the issue early tomorrow. Current weather forecasts call for few clouds at 3,000 feet and scattered clouds at 25,000 feet; visibility of 7 miles; winds from the north at 12 peaking to 20 knots; temperature at 52 degrees F; relative humidity at 69 percent; and no chance of showers. Weather officials indicate a 90 percent that weather will be favorable for a launch on Thursday. Forecasters predict an 80 percent chance of favorable weather for launch attempts on Friday and Saturday. Bruce Buckingham. (1999). **Kennedy Space Center Space Shuttle Status Report**

[Online]. Available E-mail: domo@news.ksc.nasa.gov/subscribe shuttle-status [1999, December 15].]

DECEMBER 16: Space Shuttle Status Report, Thursday, December 16, 1999. STS-103: Work in progress: Space Shuttle managers today announced a 24-hour delay in the launch of Shuttle Discovery on mission STS-103. The next launch opportunity is at 8:47 p.m. Friday, Dec. 17. Shuttle engineers have requested the additional time to confirm the quality control processes for Arrowhead Products of Los Alamos, California. Arrowhead is a vendor that provides welded propellant lines to the Shuttle Program, including main propulsion system (MPS) lines in the orbiter's aft engine compartment. On Tuesday, the vendor's quality control personnel notified NASA of a welding flaw on a pressure line being fabricated for use on a future external tank. While Shuttle engineers have already exonerated Shuttle Discovery's external tank of any welding flaws, the on going evaluation will confirm the quality of the procedures used to manufacture Shuttle MPS lines. Though managers are confident in the structural integrity of Discovery's MPS lines, this evaluation is being conducted as a precaution. Launch managers at KSC are not working any significant issues and stand ready to support tomorrow's launch opportunity. The countdown clock will hold at the T-11 hour mark and a decision to begin tanking operations will be made at about noon tomorrow. Weather officials indicate a 60 percent chance that weather will prohibit tomorrow's launch. The primary concern is thick cloud layers and the possibility of coastal showers. Bruce Buckingham. (1999). **Kennedy Space Center Space Shuttle Status Report** [Online]. Available E-mail: domo@news.ksc.nasa.gov/subscribe shuttle-status [1999, December 16].]

◆ Boaters and fishermen who are used to seeing weather buoys in the ocean east of Cape Canaveral may wonder at the unusual devices on new ones recently placed there. Those devices are vital in relaying hourly weather data, especially during a launch countdown and landing. The closest is 20 nautical miles offshore from the Cape in the Gulfstream; another is a distant 110 nautical miles ENE. The National Weather Service in Melbourne also uses the data for accurate forecasts of wind and seas over the area offshore from Brevard County. ["Data from weather buoys critical to launch countdown," **KSC Countdown**, December 16, 1999.]

◆ The launch of a NASA satellite that will study Earth's environment was automatically halted by a computer moments before liftoff Thursday. It was not immediately known what caused the cutoff. The launch of the Terra satellite aboard an Atlas IIAS rocket was called off for the day. NASA said it will probably try again in 24 hours. ["Computer stops liftoff of $1.3 billion satellite," **Florida Today**, December 17, 1999, p 2B.]

DECEMBER 17: Space Shuttle Status Report, Friday, December 17, 1999. STS-103: Work in progress: Loading of Shuttle Discovery's external tank began at about 11:49 a.m. today and concluded at about 2:45 p.m. With Discovery in excellent health and the launch team ready to support tonight's 8:47 p.m. launch attempt, Shuttle managers decided to proceed with tanking operations during a standard pre-tanking meeting this morning. During the mission management team meeting, Shuttle engineers presented the results of their recent quality control review of vendor Arrowhead Products. After thorough evaluation of Arrowhead's production history and quality control system, managers gained increased confidence in the structural integrity of Discovery's main propulsion system lines and confirmed that the single welding flaw identified earlier this week was an isolated incident. The precautionary review also confirmed arrowhead's high level of quality workmanship. The weather forecast calls for clouds to be broken at 3,000 feet and 5,000 feet, and overcast at 12,000 feet; visibility at 5 miles; winds out of the north east at 12 knots gusting to 20 knots; temperature at 69 degrees F; relative humidity at 88 percent and rain showers in the KSC vicinity. Tonight, forecasters indicate only a 20 percent chance that weather will be acceptable at launch time. The primary concerns are thick layered clouds and rain showers. The 24-hour delay forecast improves to a 40 percent chance of favorable weather for a Saturday launch attempt at 8:21 p.m. EST. Bruce Buckingham. (1999). **Kennedy Space Center Space Shuttle Status Report** [Online]. Available E-mail: domo@news.ksc.nasa.gov/subscribe shuttle-status [1999, December 17].]

◆ Updated Space Shuttle Status Report, Friday, December 17, 1999. STS-103: Space Shuttle mission managers scrubbed the launch of Discovery on mission STS-103 tonight at about 8:52 p.m. due to numerous violations of weather launch commit criteria at KSC. A 24-hour scrub turnaround sequence is currently in work at the pad. Managers will attempt to launch Discovery tomorrow, Dec. 18, at 8:21 p.m. Eastern Time. The 42-minute window extends to 9:03 p.m. Currently at the pad, operations to detank the external tank of its unused liquid propellants will continue into the early morning. Following this operation, Discovery will then be placed in a posture to be refilled with the cryogenics beginning around 11:30 a.m. The seven-member crew of Discovery have returned to their crew quarters for the night and will begin their sleep period about 2:30 a.m. They will awake tomorrow at about 10:30 a.m. and once again go through their launch day routine. Walkout of the Operations and Checkout Building is scheduled for 4:47 p.m. Tomorrow's weather outlook is somewhat similar to tonight's. Forecasters are predicting a 70 percent chance of weather criteria violation with the primary concerns being for low clouds and possible rain showers in the vicinity of the pad and the Shuttle Landing Facility. Bruce Buckingham. (1999). **Kennedy Space Center Space Shuttle Status Report** [Online]. Available E-mail: domo@news.ksc.nasa.gov/subscribe shuttle-status [1999, December 17].]

DECEMBER 18: Space Shuttle Status Report, Saturday, December 18, 1999. STS-103: Work in progress: Today, Shuttle managers decided to delay Discovery's launch due to the increased threat of unfavorable weather. Launch managers intended to begin operations to load the external tank at 11 a.m. today, but with a 70 percent chance of weather violation they delayed a go ahead decision to collect more weather data. By noon, weather officials reported an 80 percent probability that weather would prohibit tonight's launch attempt. The primary concerns are thick layered clouds and rain showers. Shuttle managers are assessing the possibility of launching Discovery on Sunday, Dec. 19. Current forecasts indicate a 60 percent chance of favorable weather. Along with weather, managers are reviewing the feasibility of supporting contingency landing operations at Edward Air Force Base, CA, prior to the new year. Tomorrow's forecast calls for clouds to be scattered to broken at 3,000 feet, broken at 7,000 feet, and overcast at 12,000-22,000 feet; visibility at 7 miles; winds out of the north east at 12 knots gusting to 20 knots; temperature at 69 degrees F; and rain showers in the KSC vicinity. Bruce Buckingham. (1999). **Kennedy Space Center Space Shuttle Status Report** [Online]. Available E-mail: domo@news.ksc.nasa.gov/subscribe shuttle-status [1999, December 18].]

◆ A NASA satellite was launched Saturday on a $1.3 billion-plus mission to observe the interactions among Earth's land masses, atmosphere, ocean and biosphere. The Terra satellite, carrying five sophisticated instruments, lifted off at 1:57 p.m. atop a Lockheed Martin Atlas IIAS rocket. ["Earth-observing satellite lifts off in California," **Florida Today**, December 19, 1999, p 1A.]

DECEMBER 19: Space Shuttle Status Report, Sunday, December 19, 1999. STS-103: Work in progress: Today, Shuttle managers decided to proceed with the STS-103 launch countdown and loading Discovery's external tank. The tanking activity began at 10:30 a.m. today and will take approximately 3 hours to complete. The countdown clock will enter a planned 2-hour built-in hold at the T-3 hour mark at 2 p.m. The clock will resume at 4 p.m. and enter the next planned built-in hold at the T-20 minute mark at 6:40 p.m. for 10 minutes. The final planned hold will come at the T-9 minute mark at 7:01 p.m. for 40 minutes. The clock will resume at 7:41 p.m. for an on-time launch at 7:50 p.m. Tonight's weather forecast calls for improved conditions as the weak frontal boundary continues to drift north away from Central Florida. There is an 80 percent chance for acceptable weather conditions tonight. Clouds will be scattered at 3,000 feet, 8,000 feet, and 25,000 feet; visibility will be 7 miles; winds out of the southeast at 10 knots; temperature at 65 degrees F; and there is a slight chance of a low cloud ceiling. Although the Kennedy Space Center is the prime end-of-mission landing site, Shuttle managers carefully reviewed turnaround-processing operations at Edwards Air Force Base, CA, should landing be diverted. There are several landing opportunities in both Florida and California on the planned landing on Dec. 27. Historical weather data indicates both sites have a high probability of favorable weather this time of year at the planned landing time. Bruce Buckingham. (1999). **Kennedy Space Center Space Shuttle Status Report** [Online]. Available E-mail: domo@news.ksc.nasa.gov/subscribe shuttle-status [1999, December 19].]

◆ Updated Space Shuttle Status Report, Sunday, December 19, 1999. STS-103: Work in progress: Space Shuttle Discovery and a seven-member flight crew lifted off from KSC's Launch Pad 39B on time at 7:50:00:069 p.m. today. The launch team worked no significant issues during the launch countdown and weather conditions were excellent at launch time. Discovery has embarked on its 27th space flight. Discovery and crew will rendezvous with the Hubble Space Telescope on Tuesday afternoon for a berthing of the spacecraft in the orbiter's payload bay Tuesday night. Three space walks are planned during this mission to accomplish planned Hubble servicing efforts. Discovery returns to earth Monday, Dec. 27 at about 5:24 p.m. EST. The solid rocket booster recovery ships, Liberty Star and Freedom Star, deployed from KSC on Wednesday, Dec. 15. They are expected to arrive at Hangar AF with boosters in tow tomorrow at about 4:30 p.m. EST. Bruce Buckingham. (1999). **Kennedy Space Center Space Shuttle Status Report** [Online]. Available E-mail: domo@news.ksc.nasa.gov/subscribe shuttle-status [1999, December 19].]

DECEMBER 20: Space Shuttle Status Report, Monday, December 20, 1999. STS-103: Work in progress: Space Shuttle Discovery continues to perform very well on orbit as the flight crew prepares to rendezvous with, grapple and service the Hubble Space Telescope. Following Discovery's launch last night, workers completed a preliminary walk down of Launch Pad 39B. Inspectors report only minimal damage at the pad. Just before noon today, the solid rocket booster recovery ships began towing operations and are expected to arrive at Hangar AF Tuesday at about 4:30 a.m. Both boosters are reported to be in good condition following last night's splashdown in the Atlantic Ocean. Bruce Buckingham. (1999). **Kennedy Space Center Space Shuttle Status Report** [Online]. Available E-mail: domo@news.ksc.nasa.gov/subscribe shuttle-status [1999, December 20].]

◆ Kennedy Space Center will reorganize its work force by April to fill "critical skill shortages" that have developed from NASA staff cuts in recent years, Space Center Roy Bridges said Monday. Highlights of the plan include hiring 75 additional engineers and technicians, reducing the number of top managers and shifting an unspecified number of workers into new jobs and duties. The changes are partly aimed at fulfilling the Space Center's "commitments for safe operations" in the space shuttle, International Space Station and Expendable Launch Vehicle programs, according to the plan. Other goals of the reorganization include improving NASA's expertise and aligning the agency to "live within available resources," the plan sates. Richard Blomberg, a member of the Aerospace Safety Advisory Panel that serves as a NASA watchdog group for Congress, expressed support. "They recognize the issues and are taking steps in the right direction," Blomberg said of Bridges' announcement. "Certainly, bringing in additional engineering talent is a step in the right direction." Problems with the Space Center's organizational structure include unnecessary duplication of jobs and diluted authority and accountability, the plan states. The reorganization will be the largest at the Space Center in more than 15 years, Bridges said. He said the center's current organization is based on a larger work force and is no longer suitable. Previously formulated plans call for reducing the number of civil-service employees at the center from 1,665 this year to 1,422 in 2008. But Bridges said the reorganization announced Monday is not being done to further reduce the work force. "This is not about downsizing. This is about positioning ourselves for the future," he told about 275 Space Center workers attending a private presentation of the plan televised on closed-circuit television. ["KSC plans work force overhaul," **Florida Today**, December 21, 1999, p 1A & 2A.]

DECEMBER 27: Space Shuttle Status Report, Monday, December 27, 1999. The orbiter Discovery landed successfully tonight on the second KSC landing opportunity at 7:01 p.m. EST. The first landing opportunity was waved-off due to unacceptably high cross winds at the Shuttle Landing Facility. The landing occurred on KSC Shuttle Landing Facility (SLF) runway 33. On this mission, STS-103, the orbiter and crew traveled over 3,267,000 miles.
End of mission elapsed times are:

	Eastern Time	Mission Elapse Time
Main Gear Touchdown minutes/47 seconds	7:00:47 p.m.	7 days/23 hours/10
Nose Gear Touchdown minutes/58 seconds	7:00:58 p.m.	7 days/23 hours/10
Wheels Stop minutes/34 seconds	7:01:34 p.m.	7 days/23 hours/11

Upon close inspection of the orbiter following touchdown, engineers noted that a black tile was missing on the right inboard elevon, next to the fuselage. The missing tile measures 9 inches by 41/2 inches. No significant damage to the orbiter was found and the flight crew was never in any danger due to the missing tile. Initial indications are the tile came off sometime just prior to final approach. Further analysis will take place over the next several days once the orbiter is in the Orbiter Processing Facility. Tow to the OPF is scheduled to begin at about 12 midnight tonight. Once in the OPF, the Discovery's systems will be deserviced and vehicle safing will be conducted over the next two days. Discovery, as well the other two vehicles at KSC, will be powered-down for the remainder of the holidays. Normal vehicle processing is scheduled to commence Jan. 4. The seven-member astronaut crew will spend tonight in Florida. Tomorrow they are scheduled to depart at about 2:30 p.m. from Patrick Air Force Base for their homes in Houston, TX. Bruce Buckingham. (1999). **Kennedy Space Center Space Shuttle Status Report** [Online]. Available E-mail: domo@news.ksc.nasa.gov/subscribe shuttle-status [1999, December 27].]

DECEMBER 28: After an eight-day mission to repair the Hubble Space Telescope and an evening landing Monday at the Kennedy Space Center, shuttle Discovery was towed Tuesday to its hangar where NASA officials hope it will now be safe from the Y2K bug. KSC workers spent the day draining hazardous fluids from Discovery and by early this evening will have the effort completed and the plug pulled on the spaceship's power supply, KSC spokesman George Diller said Tuesday. "The intent is to safe and stabilize the orbiter so that we can power it down through the New Year, Y2K period," Diller said. Discovery's launch and landing was dictated by NASA's desire to have the shuttle on the ground and turned off by the end of today. The fact that most space center workers are off this week because o the holidays is helping make Discovery's hazardous work go faster. That's good news to the 30 technicians and engineers working in shifts around the clock to make Discovery safe, because the sooner they can get the work done, the sooner they can enjoy the rest of the week off. Early inspections show Discovery held up well during its 27th spaceflight. One problem: a black heat protection tile measuring nine inches long and 4.5 inches wide was found missing from an orbiter body flap. From the lack of heat damage to the underlying orbiter skin, it is likely the tile fell off while the shuttle was making its final approach to the space center runway, Diller said. ["Discovery safely back in hangar in time for 2000," Florida Today, December 29, 1999, p 1B.]

REPORT DOCUMENTATION PAGE

1. AGENCY USE ONLY (Leave blank)	2. REPORT DATE	3. REPORT TYPE AND DATES COVERED
	February 2000	Technical Memorandum 1999

4. TITLE AND SUBTITLE

Chronology of KSC and KSC Related Events for 1999

5. FUNDING NUMBERS

6. AUTHOR(S)

Elaine E. Liston

7. PERFORMING ORGANIZATION NAME(S) AND ADDRESS(ES)

InDyne, Inc.
Kennedy Space Center

8. PERFORMING ORGANIZATION REPORT NUMBER

9. SPONSORING/MONITORING AGENCY NAME(S) AND ADDRESS(ES)

10. SPONSORING/MONITORING AGENCY REPORT NUMBER

NASA/TM-2000-208588

11. SUPPLEMENTARY NOTES

This chronology is published to fulfill the requirements of KMI 2240.1 (as revised) to describe and document KSC's role in NASA's progress.

12a. DISTRIBUTION AVAILABILITY STATEMENT

12b. DISTRIBUTION CODE

13. ABSTRACT (Maximum 200 words)

This document is intended to serve as a record of KSC events and is a reference source for historians and other researchers. Arrangement is by day and month and individual articles are attributed to published sources. Materials were researched and described by the KSC Library Archivist for KSC Library Services Contractor InDyne, Inc.

14. SUBJECT TERMS

Chronology, Space Shuttle Orbiters, Expendable Launch Vehicles Chandra X-Ray Observatory, Safety, Kennedy Space Center Visitor Complex, Payloads, Bridges, Launches, Landings

15. NUMBER OF PAGES

147

16. PRICE CODE

17. SECURITY CLASSIFICATION OF REPORT	18. SECURITY CLASSIFICATION OF THIS PAGE	19. SECURITY CLASSIFICATION OF ABSTRACT	20. LIMITATION OF ABSTRACT
Unclassified	Unclassified	Unclassified	

www.ingramcontent.com/pod-product-compliance
Lightning Source LLC
Chambersburg PA
CBHW080252180526
45167CB00006B/2497